T0212435

Synthese Library

Studies in Epistemology, Logic, Methodology, and Philosophy of Science

Volume 423

Editor-in-Chief
Otávio Bueno, Department of Philosophy, University of Miami, Coral Gables, USA

Editors
Berit Brogaard, University of Miami, Coral Gables, USA
Anjan Chakravartty, University of Notre Dame, Notre Dame, USA
Steven French, University of Leeds, Leeds, UK
Catarina Dutilh Novaes, VU Amsterdam, Amsterdam, The Netherlands

The aim of *Synthese Library* is to provide a forum for the best current work in the methodology and philosophy of science and in epistemology. A wide variety of different approaches have traditionally been represented in the Library, and every effort is made to maintain this variety, not for its own sake, but because we believe that there are many fruitful and illuminating approaches to the philosophy of science and related disciplines.

Special attention is paid to methodological studies which illustrate the interplay of empirical and philosophical viewpoints and to contributions to the formal (logical, set-theoretical, mathematical, information-theoretical, decision-theoretical, etc.) methodology of empirical sciences. Likewise, the applications of logical methods to epistemology as well as philosophically and methodologically relevant studies in logic are strongly encouraged. The emphasis on logic will be tempered by interest in the psychological, historical, and sociological aspects of science.

Besides monographs *Synthese Library* publishes thematically unified anthologies and edited volumes with a well-defined topical focus inside the aim and scope of the book series. The contributions in the volumes are expected to be focused and structurally organized in accordance with the central theme(s), and should be tied together by an extensive editorial introduction or set of introductions if the volume is divided into parts. An extensive bibliography and index are mandatory.

More information about this series at http://www.springer.com/series/6607

Paul Needham

Getting to Know the World Scientifically

An Objective View

Springer

Paul Needham
Department of Philosophy
University of Stockholm
Stockholm, Sweden

Synthese Library
ISBN 978-3-030-40218-1   ISBN 978-3-030-40216-7   (eBook)
https://doi.org/10.1007/978-3-030-40216-7

Preface

Scientific knowledge is nothing if not objective. Men are not flown to the moon on the basis of the whim of any Tom, Dick or Harry, or even the president of the United States, however much they might wish it. No such project would have been feasible prior to WW2 but was made possible by the application of sound scientific knowledge built up over the generations and the well-established expertise of implementing it in engineering practice. Such engineering feats and the acquisition of scientific knowledge on which they depend are fundamentally social phenomena, relying on the accumulation of knowledge and what Hilary Putnam has called the sharing of the linguistic burden. The wheel has not been reinvented with each new generation, which has not had to begin from scratch. Knowledge has been passed on from generation to generation, modified, elaborated and supplemented in the process as demanded by the ever-expanding experience by scientists standing, as Newton famously put it, on the shoulders of others. As the body of accumulated knowledge expanded beyond the grasp of any one individual, surpassing the likes of the so-called universal minds such as Leibniz and Newton who commanded substantial proportions of the available knowledge of their day but became a phenomenon of bygone centuries, experts from different fields cooperated in tackling the increasingly complex issues that the scientific community addressed. Research teams typically comprise experts in various fields of specialisation, each drawing on others' knowledge which they respect and comprehend to the extent of being able to understand its role in the overall investigation. And so it is in modern, everyday life where we are prepared to grant that there are, for example, chalk hill blue butterflies that someone more knowledgeable than ourselves could identify and classify even if we can't do it ourselves or oaks that are distinguished as pedunculate or sessile even if we haven't the faintest idea of what marks the distinction and why.

Despite the many ostentatious demonstrations of the achievement of scientific knowledge, the social aspect of science and complexity of its projects has led critics to question its objective status in the name of social constructivism, postmodernism and relativism. The critique is often directed at a straw man, a once-influential view that is long since outdated, or a misrepresentation of a more interesting view: the world consists of a countable jumble of facts like a collection of lego pieces in a box, the logical positivists' doctrine that scientific progress amounts to the accumulation of more and more observations, the early Wittgenstein's picture theory of truth, and so on. Often, the case presented amounts to a non sequitur. The fact that knowledge claims are formulated in terms of concepts that emerged from a historical process of deliberation doesn't imply that the claims themselves are "social constructions" rather than what is true of the world. We use our concepts to describe the world. It might well be felt that the critique of objectivity can be safely ignored as the outpourings of a group of uninteresting sociologists, as many scientists have done, were it not for the disastrous influence it has had on secondary school curricula and teaching practice.[1] At all events, I won't be trawling through the social constructivism literature here. Rather, the plan is to give some account of how the social dimension of science underlies its objectivity, and to take up and criticise some lines of thought underlying relativism, which is one of the main sources of inspiration for social constructivism.

Relativism is associated with antirealism—one of the classical issues in the philosophy of science. Antirealism postulates that science is concerned with the systematic recording of observations, capturing regularities in the concomitance of the observable features of common-sense bodies with a view to predicting the occurrence of observable features in the future. This calls for efficient unifying organisation, linking apparently unconnected observable features, which is promoted by the postulating of unobservable entities and their associated features. These provide a theoretical foundation in terms of which the observational regularities can be simply and efficiently encompassed in a unified system. On this understanding of the observational and theoretical parts of the body of scientific knowledge, the theoretical core might be removed once its unifying aid has served its heuristic purpose of bringing the observational regularities into place without detriment to the overall truth of the body of knowledge. For, according to the antirealist, whatever evidence supports this body of knowledge is based on observation, and that merely supports the observational part. Traditional empiricists following Berkeley have held some such view, maintaining that observations concern the private sense data experienced by individual minds, which is what constitutes observational knowledge. This is idealism, according to which even the everyday, common-sense objects around us are theoretical constructions facilitating the organisation of this knowledge although in principle redundant. But sustained critique had extinguished the last vestiges of the sense data-based notion of observation by the mid-twentieth century. The logical positivists of the 1920s and 1930s construed observational knowledge in terms of intersubjectively agreed perception. That fell by the wayside in the 1950s when the theory dependence of observation was generally recognised (although the idea goes back to Duhem at the turn of the century). Antirealists, who bear the onus of delimiting an appropriate notion of

---

[1] The pernicious influence of social constructivism on Swedish schools, which have seen a marked decline in standards as recorded by international comparisons since the mid-1990s, has been well documented by Åsa Wikforss (2017).

observational knowledge, have subsequently taken some steps to accommodate theory dependence in their conception of observation. But there is no support for such projects in scientific practice, where observation is not understood to delimit a more directly accessible domain of nature. The realist response advocated here, which with some qualifications affirms the literal truth of scientific theories, follows Duhem's understanding of experimental practice as not involving any such distinction between the accessible and inaccessible domains of nature and allies it with Duhem's historical thesis of the continuity of science.

These broad themes are developed in what is intended as an introduction to some fundamental issues in the philosophy of science, which I hope will be of value both for students of philosophy and science. Part I deals with knowledge and values. Chapter 1 starts the ball rolling with a presentation of the classical conception of knowledge as initiated by the ancient Greeks and elaborated during the development of science, introducing the central concepts of truth, belief and justification. Aspects of the quest for objectivity in claims to know are taken up in the following two chapters. Justification is discussed in Chap. 2, rounding off with a discussion of the interplay of values and statistics in scientific inference, and the objective claims of truth are taken up in Chap. 3. Moral issues are broached in Chap. 4 which discusses some aspects of the use and abuse of science, taking up the responsibilities of scientists in properly conducting their business and decision-makers in their concerns with the import of science on society.

Part II looks at some philosophies of science. Some philosophers see the progress of science primarily in terms of rejecting old hypotheses and theories and replacing them with new ones, whereas others see science as progressing primarily by accumulating knowledge, saving as much as possible from older theories in the course of developing new theories accommodating new experimental and observational results. The Austrian philosopher Karl Popper, discussed in Chap. 5, is a well-known representative of the first group, and the French physicist, historian and philosopher Pierre Duhem, a well-known representative of the second group, is taken up in Chap. 6. A concluding chapter discusses the natural attitude of taking the theories of modern science to be literally true, i.e. realism, in the light of arguments drawn from the history of scientific progress in favour and in criticism of this stance.

Some points in the first chapters may already be familiar to some students. But I hope this will be mitigated by the general context of the presentation by placing epistemological issues raised in philosophy courses in a scientific context and relating experimental procedures to epistemology. Moral issues are raised in Chap. 4 and some of what is said there might be thought controversial. But this hopefully raises important issues and serves to stimulate reflection and guide discussion. The discussion of positivism in Chap. 5, largely on the basis of Ayer's understanding, may seem overly simplistic in the light of the recent interest in the detailed history of the development of logical positivism. But this is not the place to go into a more rigorous historical treatment, bringing to light the subtlety of thought of figures such as Carnap and Reichenbach. I have tried to illustrate points with examples taken from contemporary or historical science rather than leaving them in the abstract, but without pursuing details to the extent of confounding readers unfamiliar with the relevant specialised knowledge. References often provide or lead to further information. I have tried on the whole to avoid technical matters of logic, although it is sometimes appropriate to mention where they enter the debate. Again, references point to way for those interested in pursuing such matters.

This book is largely based on a course in the philosophy of science that I have given over the years at Stockholm University for a mixed group of students and occasionally staff from other departments, including some who were reading or had read philosophy and others reading sciences, engineering, humanities and social sciences. I am grateful for the interest and enthusiasm evident from their questioning, which has had a considerable influence on the final form of this book.

Stockholm, Sweden   Paul Needham
April 2017

Contents

Part I
Knowledge, Objectivity and Values

Chapter 1
Knowledge

1.1 Ideals Deriving From the Greeks

According to one traditional view of the history of science, there were two important stages in the development of the modern conception of science. The first awakening came with the efforts of the ancient Greek philosophers to free themselves from the mythical stories of how the world is governed by the wills of the Gods and build a rational and systematic body of genuine knowledge. But Greek science descended into the dark ages, when medieval mysticism impeded and delayed the progress of true science. This state of affairs ruled until the rejection of medieval mysticism in the so-called scientific revolution in the sixteenth and seventeenth centuries, which constituted the second stage when the foundations for modern empirical science were laid. As we will see in Chap. 6, this view greatly oversimplifies the historical development of science. But the point remains that the ancient sources have exerted a considerable influence on the acquisition and understanding of knowledge.

The classical ideal of the Greeks sought to base all genuine knowledge, which was to provide complete explanations and final answers to our questions, in self-evident truths. These were to be discerned by systematic reflection on the myriads of opinions about the world around us and formulated as axioms from which all other knowledge could be deduced by the strict application of valid principles of logic. Euclidean geometry corrected and systematised in this way the sundry facts about the properties of areas and solid bodies which earlier cultures had discovered on the basis of observation and the method of trial and error. Thales (c. 625–545 BC), the earliest of the Greek philosophers known to us, is said to have been the first to actually formulate a proof of one geometrical proposition from others, and the edifice formed by those who built on his insight was eventually put together in Euclid's *Elements* (c. 300 BC). Aristotle, and later the Stoics, set about explicitly formulating the universal principles of logic on which all reasoning, including that used to derive results in geometry, is based.

© Springer Nature Switzerland AG 2020
P. Needham, *Getting to Know the World Scientifically*, Synthese Library 423,
https://doi.org/10.1007/978-3-030-40216-7_1

This made possible the discovery of new geometrical truths as never before, not by directly observing spatial features of the world, but by demonstrating theorems which strictly followed from the axioms by the principles of logic. But why believe them true? Two aspects just mentioned of this way of organising knowledge were designed to address this question. The principles of logic guarantee that inferences preserve truth. That is to say, *if* the premises at each stage of the argument are true, *then* the conclusion is true. An argument proceeding by stages of this kind would thus never lead to a falsehood *provided* it doesn't begin with a falsehood. Arguing strictly in accordance with logic is therefore not enough. We must also be assured that the starting points of the argument (i.e. propositions not themselves proved by appeal to others) are true. These starting points are the axioms, which are acceptable as such if they are propositions of such simplicity that their truth is self-evident and could not be doubted. The axioms of Euclidean geometry were held to express necessary truths—truths which couldn't possibly be false. Even this necessary character was supposed to be conveyed to their consequences by the principles of logic. But the self-evidence may be difficult to ascertain in a complex theorem. This is why, in order to be reliable, the complexity must be built up from self-evident starting points in short, reliable steps.

Aristotle (384–322 BC) held that the axiomatic systematisation of geometry, which was well under way in his time, provided the model on which all knowledge should be built. Genuine knowledge provides an insight into nature which he called understanding, and characterised as follows:

> We think we understand a thing simpliciter (and not in the sophistic fashion accidentally) whenever we think we are aware both that the explanation because of which the object is its explanation, and that it is not possible for this to be otherwise.
>
> ... it is necessary for demonstrative understanding in particular to depend on things which are true and primitive and immediate and more familiar than and prior to and explanatory of the conclusion. ... there will be deduction even without these conditions, but there will not be demonstration; for it will not produce understanding.
>
> Now they [i.e. these "things which are true and primitive and ..."] must be true because one cannot understand what is not the case .... (*Posterior Analytics*, 1.2; 71^b9–25)[1]

Philosophers after Aristotle have been virtually unanimous in accepting the basic principle that knowledge is only possible of what is true—knowledge, in a word, implies truth. As we will see, this has been challenged by relativists, but to no avail. Other aspects of this ancient conception of knowledge—and in particular the need to fall back on what cannot be otherwise, i.e. necessary truths—have been less hardy. Nevertheless, medieval thinkers would argue from what they perceived to be necessary truths. We still find major thinkers labouring under a conception of knowledge based on necessary principles at the turn of the seventeenth century, and it is interesting to consider why they were drawn to this classic account.

---

[1]Quotations from Aristotle are taken from the English translations of his works in Aristotle (1984).

Actually, these pioneering seventeenth-century thinkers considered themselves to be doing something radically different from their predecessors and revolutionising science. But this wouldn't work if they didn't adhere to some common ground rules, in particular those determining what is to count as knowledge.

Galileo (1564–1642) described himself as endeavouring "to investigate the true constitution of the universe—the most important and most admirable problem that there is", which clearly shows him to be still in the grip of the Greek tradition. "For", he continues, "such a constitution exists; it is unique, true, real, and could not possibly be otherwise" (1613, p. 97). He was quite sure that "the natural sciences, whose conclusions are true and necessary ... have nothing to do with human will" (1632, p. 53). Indeed, "with regard to those few [propositions] which the human intellect does understand, I believe that its knowledge equals the Divine in objective certainty, for here it succeeds in understanding necessity, beyond which there can be no greater sureness" (1632, p. 103). Science is objective, Galileo says, and so independent of our desires. He seems to have thought that what is necessarily true would satisfy this requirement of objectivity, since whatever is necessary is not only true, but beyond the power of human influence and thus not subject to human will.

If objectivity is to be claimed for necessary truth, then necessity must be established by appeal to insights or principles available to whole communities and not convictions peculiar to a particular person. It would be a discursive matter, in which truth is established by propounding an argument which an opponent would find irresistible, starting from a common ground of agreed premises. This is how Galileo conceived his task. He presented his case for the heliocentric system Copernicus (1473–1543) had presented almost a century earlier in the form of a dialogue. There are three interlocutors in his *Dialogues Concerning Two Chief World Systems* (1632). Salviati, who speaks for Galileo, disputes with Simplicius, named after the sixth century Aristotelian commentator and representing Aristotle, in the presence of a third interlocutor, Sagredo, who represents the voice of unprejudiced common sense. But the pioneers of seventeenth-century science were constrained by their opponents. Galileo, obliged by a decree of the Church in 1616 to treat the Copernican thesis as a hypothesis and not to profess it as a truth, was also bound by the Augustinian doctrine.[2] This laid it down that literal interpretation of the Bible, which asserts in many places that the earth is stationary and the sun moves, should be preferred to bare assertions to the contrary, and should only be relinquished where the contrary is demonstrated.

Galileo tried to sidetrack the authority of received opinion and appeal directly to common sense by writing in Italian, avoiding the mindless jargon associated with the standard academic medium of Latin with its ingrained formulas and categories. In the same vein, Descartes (1596–1650) says in his 1637 *Discourse on Method*, "And if I am writing in French, my native language, rather than in Latin, the language of my teachers, it is because I expect that those who use only their

---

[2]St. Augustine (354–430), a key figure in the transition from classical antiquity to the Middle Ages, is noted for his adaptation of classical thought to Christian teaching.

natural reason in all its purity will be better judges of my opinions than those who give credence only to the writings of the ancients" (VI, p. 151). But he was distinctly less optimistic than Galileo of finding common ground with his opponents, suggesting that

> it is custom and example that persuade us, rather than any certain knowledge. And yet a majority vote is worthless as a proof of truths that are at all difficult to discover; for a single man is much more likely to hit upon them than a group of people. I was, then, unable to choose anyone whose opinions struck me as preferable to those of all others, and I found myself as it were forced to become my own guide. (Descartes 1637, p. 119)

Descartes advocated a method by which he tested and relied on his own intellectual insight or intuition, pitting his own resources against received opinion by focusing on the clear and distinct perception of simple things and their necessary connections. Accordingly, he wrote in the first person, enticing the reader to follow his example and think for himself. The reader should arrive at his own conclusions by adopting a sceptical attitude towards received opinion until he could establish the matter with certainty for himself. This was the method of doubt, of systematically searching for any grounds for doubting a proposition, which Descartes would refuse to accept until he was satisfied that he had good reason to think it was beyond doubt. His approach to the problem of establishing genuine knowledge was thus to become certain; whatever fell short of certainty fell short of being knowledge.

The extent to which his quest for certainty and the elimination of doubt led him to rely on his own intellectual resources in assessing his claim to knowledge is why he is said to belong to the rationalist school. Rationalists were impressed by the reasoning and insight involved in acquiring the new knowledge accruing in the scientific revolution, which often led to the appreciation that things were not as they seemed to be. The sun seems to rise in the morning and set in the evening, so revolving around the earth. But the Copernican system, which revolutionised our view of the heavens by claiming that the earth revolves around the sun, was defended by Galileo and Descartes as the true account of the universe. This entailed that the retrograde motions of the planets observed by the Greeks and incorporated in the ancient systems surviving through the renaissance, were an illusion. As observed against the backdrop of the fixed stars, each planet's motion follows a general pattern in which a steady progression to the east is intermittently interrupted by periods of retrograde motion during which the planet returns on its path and moves westwards. These irregular motions were incorporated literally into the ancient geocentric system by introducing a complicated system of epicycles. According to the Copernican system, however, the retrograde movements are not literally executed by the planets, but are just appearances arising in virtue of the observer's motion relative to the planets. With some further help from Kepler (1571–1630), it was possible to remove epicycles on the basis of this insight and obtain a much more simple and elegant account of the solar system. Again, in defending the Copernican system against the arguments which seem to speak in favour of a stationary earth, Galileo's analysis of falling bodies introduced motions that are not directly observed and appear not to be executed by falling bodies. In this spirit, he couldn't withhold his admiration, despite "the experiences which overtly contradict the annual movement", for "Aristarchus and Copernicus [who] were able to make reason so conquer sense that in defiance of the latter, the former became mistress of their belief" (Galileo 1632, p. 328).[3] The concepts and principles which really guide and form our scientific knowledge, the rationalists thought, derive not from experience, but have their origin in the intellect and are innate in the human faculty of thinking.

A great epistemological debate raged in the seventeenth and early eighteenth centuries between the rationalists and members of the opposing empiricist school, who, while allowing that there were truths of reason, thought that much knowledge is ultimately based on experience and observation. Knowledge, they agreed with the rationalists, must rest on indubitable foundations. But these foundations include direct experience of the world. To ensure certainty, however, the sensory experience forming the basis of our empirical knowledge would have to be distinguished from idiosyncrasies arising as a result of misperception, illusion or hallucination. Could this be done without appealing to reason and invoking the feature of necessary truths which would make them subservient to the intellect as the rationalists thought? The empiricists launched a response on two fronts, maintaining certainty by making ever more guarded claims about what the content of our sensory experience actually is, and ridiculing the notion of innate knowledge.

These traditional views, which have it that knowledge is restricted to what is absolutely certain and suggest that we know very little indeed, haven't prevailed. Most of us believe today that we can acquire knowledge of contingent matters of fact—what ordinary experience and the empirical sciences teach us without an absolute guarantee of certainty. Kant (1724–1804), writing towards the end of the eighteenth century and traditionally described as compromising between the extremes of rationalism and empiricism, presented a more balanced view, more in tune with our present conception of what we know. Accepting that we have knowledge of contingent facts doesn't automatically mean that we have no knowledge at all of things like the necessary truths which the Greeks took to be paradigmatic of what we know, and Kant certainly didn't wish to deny this. On the contrary, he took such knowledge to be an essential feature of our having any knowledge at all, and many philosophers after him were to go along with this, even if disagreements arise with the more detailed development of the idea. He called it a priori knowledge, which he distinguished from a posteriori, or empirical, knowledge.

The a priori truths he understood as those which are necessary, which cannot possibly be false, and the a posteriori or empirical truths as those which are contingent. But now if we are aware of contingent, a posteriori truths by virtue of observation via the senses, it might seem to be a problem how we come by a priori

---

[3] Aristarchus (c.310–c.230 BC), called "the first Copernicus", was a Greek astronomer who proposed a heliocentric account of the solar system, in opposition to the geocentric account favoured by most of his contemporaries and successors until the end of the renaissance.

knowledge. The notion that we have innate knowledge is found in Plato (428–348 BC), who thought of it as extending by analogy the fact that some of what we know is what we remember. Descartes and the seventeenth-century rationalists adopted the notion as a way of answering the question Where does our a priori knowledge come from? The rationalists were convinced that many of our ideas—the having of which is an essential prerequisite for, if not the entire content of, thinking—do not derive from experience. On the contrary, even the ability to recognise sense data as comprising ideas of such-and-such a kind presupposes, on this view, a priori knowledge. Identifying a present sense impression as one of green, for example, involves a comparison with a prototype of green whose origin cannot itself be sense experience. And if these ideas don't derive from sense impressions, they must surely be innate. But if ideas of number, geometry, the intrinsic dispositions of matter, and even God, are innate, Why, the empiricists chided, don't young children show any sign of being aware of them?

Kant criticised the whole approach, common to rationalist and empiricist alike, of construing the problem of knowledge genetically as a question of where our ideas come from. A Kantian critique of the notion of innate knowledge is that whatever talk of innateness might establish, it certainly isn't knowledge. Suppose we are born with an innate tendency to believe that $2 + 2 = 4$. Even if humans universally acknowledged this same inclination, that wouldn't itself constitute knowledge. We might just as well have been born with an innate tendency to think that $2 + 2 = 5$! On the innateness theory, it is just a happy accident that our predilection to think or say "$2 + 2 = 4$" corresponds with the truth. By distinguishing a genetic account of knowledge, in terms of its origins or causes, from the question of what counts as knowledge and how it is to be understood, Kant is drawing attention to how knowledge is justified. A priori knowledge is justified independently of all experience. How? Not everything can be demonstrated; some assumptions or basic postulates must be granted in order to show that others might be shown to follow from them. Perhaps $1 + 1 = 2$ might be taken as one such self-evident, basic postulate. There are an infinite number of similar arithmetical truths, however, and although Kant seems not to have found this peculiar, others have thought it odd that we should have to accept so many distinct such basic postulates. But it is a standing joke that in *Principia Mathematica* (1910–1912), where Whitehead and Russell set out to show that our knowledge of mathematical truths about numbers is a priori knowledge because ultimately justifiable on purely logical grounds, the authors take several hundred pages before they get to $1 + 1 = 2$. "In mathematics", these authors say on the first page of their preface, "the greatest degree of self-evidence is usually not to be found quite at the beginning, but at some later point" (p. v).

Kant talks of cognitive claims as judgements, and of having knowledge as making judgements. Whatever the psychological reality corresponding to the terminology of mental actions, judgements always involve applying concepts, i.e. general terms, and this in turn implies relating and comparing objects as falling under a given concept or not. Simple empirical judgements involve no notion of a paradigm object nor the direct apprehension of a universal, but rather the faculty of drawing certain likenesses which we learn to make. General terms don't, on Kant's view, signify universals which we directly intuit or apprehend, but stand rather for concepts whose possession is a matter of mastering a certain skill.

Rather than approaching epistemology in terms of where our ideas come from, Kant asks How is knowledge possible? What are the necessary prerequisites? He discerned an intricate structure of knowledge of the import and interrelations between concepts without the mastery of which we would never have any empirical knowledge at all. Knowledge which is in this sense prior to knowledge derived from everyday experience is a priori knowledge. But this doesn't mean a priori knowledge is temporally prior, as the innate theory would have it. The a priori status of knowledge is a question of how an item of knowledge is established as such. If no essential reference is made to any particular observations of how the world happens to be in justifying a claim to knowledge, we are dealing with a priori knowledge. This is consistent with our coming as a matter of fact to know some item of a priori knowledge—say that $2 + 2 = 4$—with the aid of observation. Perhaps we learn that $2 + 2 = 4$ by observing that combining two pairs of apples gives us four apples, and suchlike. But whether $2 + 2$ really is identical with 4 is not the sort of thing which could depend on any conceivable observation. We might wonder how we could go about justifying $2 + 2 = 4$; but if pressed to justify the claim, to say why it is that $2 + 2 = 4$, we can't point to any particular observations which do this job. Pointing to particular instances serves at best only to clarify what it is we are talking about. Perhaps we have some idea that its truth holds in virtue of deeper logical truths which sufficient reflection would show; perhaps we draw the line here and simply say that arithmetic truths are starting points, not themselves capable of justification by reduction to yet more fundamental principles but unquestionable laws of thought which just have to be accepted as such. However this might be, knowledge is a matter of how truth can be justified, and not merely the having of a representation which happens to correspond to how things really are.

1.2 What Is Knowledge?

This brief historical survey has given some indication of how the general view of knowledge has developed in recorded history. Though the historical figures canvassed above diverged on one point or another, they were all agreed that the quest for knowledge involves, first and foremost, a quest for truth. This is not to say that knowledge is the same as truth. Without worrying about what famous philosophers have said on the matter, let us try to motivate whatever other ingredients there might be.

Knowledge is not defined as truth, but on any reasonable account, knowledge implies truth. Truth is a necessary condition of knowledge, but that is not the whole story. Many truths we know nothing about. Pick some very distant star. The statement that it has no planets is true or false. If false, then the statement that it has some planets is true. But no one may even have an opinion on the matter. Did it rain on Easter Island on Christmas day, AD 800? Again, perhaps no one would venture an opinion either way. Either it did or it didn't, but if there is a question of knowledge then there must be some connection with a person. Knowledge is possessed by someone. A statement may express something that someone claims to know or not know, or perhaps someone might know something without being aware of the fact. But in any case, for something to be known there must be some connection with a person—knowledge presupposes a knower. At the very least, this requires that the person have an opinion on the matter. The true statement about whether the star just considered has a planet is not known if no one has an opinion on the matter. If we claim that some person does know something, say that the star has at least one planet, then we are claiming not only that it is true that the star has at least one planet, but also that the person believes this.

Truth and belief are necessary conditions for knowledge, but that is still not the whole story. This can be seen from the fact that opinions may differ, i.e. contradict one another. It follows, in view of the fact that knowledge implies truth, that not all of these opinions can amount to knowledge because the truth cannot be contradictory.[4] True belief is necessary, but not sufficient, for knowledge. Dogmatism may set in, and the antagonist's claims simply dismissed. But a third party, who, let us suppose, independently knows which opinion is true, will realise that mere true belief, no matter how fervently held the conviction, does not amount to knowledge, even if the belief happens to be true. Advocates of opposing views who are seriously interested in acquiring knowledge try to resolve the matter, and show that their claim to know is correct and their opponent's wrong, by offering reasons which would justify their own claim and weighing them against the opponent's reasons. The belief must be justified if what is truly believed is to count as knowledge.

This is what Galileo did when he argued that sunspots were in fact spots on, or very near, the sun. The prevailing view at the time, which would not entertain imperfections on the so-called heavenly bodies above the moon, maintained that the spots were phenomena occurring below the moon. They all apparently move across the surface of the sun in the same direction, but were said to be bodies moving, like the planets, around the earth. Because they are dark bodies, they are only visible when the sun provides the background contrast against which they can be seen. It is not possible to directly perceive depth by observation from earth, and appearances don't reveal how near or far they are from the sun. They were believed to be below the moon because that fitted with the general cosmological views inherited from the Greeks and still upheld in Galileo's day. Galileo argued on the basis of observations made with the aid of a telescope casting an image of the sun with the sunspots onto a

---

[4]An argument from Duns Scotus or one of his pupils in the fourteenth century shows that a contradiction makes it impossible to maintain a distinction between what is true and what is false. For consider any contradiction, p and not-p. From this conjunction it follows, in particular, that p. From this in turn it follows that p or r, where r is any proposition whatsoever. Returning to the original contradiction, it follows, in particular, that not-p. But this in conjunction with p or r entails r. Since this holds for any proposition, r, whatsoever, the argument shows that everything follows from a contradiction, in the face of which it is not possible to deny anything.

sheet of paper. The straightforward observation made at a single time still provided an essentially two-dimensional image which revealed no more about distance from the earth. But he gathered systematic observations of sunspots at different times during their passage over the sun's surface and argued from inferences made from this data. When sunspots first appear at the edge of the sun's disk, they are long and thin, but fatten as they progress towards the middle, after which they become gradually thinner again as they move towards the opposite edge. He also measured the speed of the sunspots across the sun's disk at various points in its trajectory over the two-dimensional image. The speed increased from the edge of the disk towards the centre, and then decreased, in a way consistent with the projection onto a two-dimensional surface of a uniform three-dimensional motion over the surface of a sphere. These findings are naturally explained by supposing that the sunspots actually move over the spherical surface of the sun, and less naturally by the supposition that the sunspots are bodies circulating the earth below the moon. But what Galileo thought clinched the matter were correlated observations showing the absence of parallax. Parallax is the apparent difference in position of an object relative to a fixed background as seen by observers at different places on the earth. Galileo corresponded with observers in northern Europe, and compared their observations of the positions of the same sunspots with his own observations of their positions made in Italy at the same time. Given the distance from the earth to the moon, it can be calculated that the distance between the place where he made his observations and that of his correspondents is sufficient for there to be a difference in the apparent positions of sunspots on their trajectory across the solar disk if they really were objects no further from the earth than the moon. Repeated comparisons showed no divergence of parallax, from which Galileo concluded that the sunspots must be considerably further from the earth than the moon, and quite near the sun.

What counts as suitable justification depends on the nature of the truth at issue. Contrast Galileo's appeal to observation with the following case.

As recorded by Diophantus, the Babylonians knew how to find integral numbers greater than 0 satisfying the equation $x^2 + y^2 = z^2$. The seventeenth-century French mathematician Pierre de Fermat (1607–1665) wrote in the margin of his copy of Diophantus' works that it is impossible to write the cube of an integer as the sum of two cubes, impossible to write the fourth power of an integer as the sum of two fourth powers, and so on, and that he had a wonderful proof that the impossibility holds for all powers greater than 2 but the small margin left him too little space to write it down. The proof, if he really had one, followed him to the grave, and the conjecture which came to be known as Fermat's last theorem became one of the most famous unsolved problems of mathematics. Mathematicians who examined the problem were unable to produce a counterexample and became convinced that the conjecture was true. But in the absence of a correct proof, their belief, however strong, was not appropriately justified, and no mathematician would claim to know that $x^n + y^n = z^n$ has no non-zero integral solutions for $n > 2$. Matters changed in the mid 1990s, when Andrew Wiles succeeded in proving Fermat's last theorem, after which it could be said to be known.

The notion of justification at issue in the claim that knowledge is justified true belief is a very broad one, ranging from empirical justification, in which the justificatory argument makes appeal to observation, to a priori justification, in which no observations are included in the reasons offered. It has the general feature of being directed towards the truth, however, and can be contrasted with other senses in which a belief might be said to be justified. An uneasy pianist, for example, afraid that a public performance might go badly because of nervousness, might be said to be justified in convincing himself that he will play well in order to gain confidence. Again, Pascal's wager recommends believing in God because if God does exist, failure to believe will incur God's wrath, whereas if God doesn't exist, the belief will incur no harm. Belief in God might then be said to be justified as the best strategy to avoid bad consequences. Beliefs might be motivated in ways such as this because they facilitate the achievement of some goal or enable someone to comfortably acquiesce in some practice or prejudice. But justifications like Pascal's wager do not motivate upholding a belief on grounds that argue for its truth, and deprives the religious believer of the intellectual self-respect that a motivation for the truth of the belief would provide.

It may seem strange to those familiar with the contrast often made between science and belief to claim that scientific knowledge entails belief. Surely the contrast implies the opposite, that they are mutually exclusive. But this is a misunderstanding of the idiom in the contrast, which should rather be compared with someone's responding to the comment "You don't like him, then" with the exclamation "I don't dislike him, I detest him!". Detesting someone entails disliking that person, but is much stronger, and a familiar idiomatic way of emphasising the stronger claim is to affirm it after denying the weaker claim. But literally speaking, detesting entails disliking; it would be a contradiction to literally claim to both detest and like someone. The idiom in question appeals to a convention in normal communication that it is misleading to make a weaker claim when a stronger claim could be made. You don't say that you dislike someone when you really detest that person because that would suggest that you merely dislike that person. Similarly, you don't say that you believe something when you know it, for that would suggest that you merely believe what is at issue. But this conversational convention doesn't imply that scientists don't believe what they claim to know. Of course they do! When nineteenth century geologists began to appreciate that the evidence meant that the earth really was much older than the 6000 odd years calculated on the basis of statements made in the bible, they didn't claim to know this and yet not believe it. Scientists who claim to know that the earth is much older than 6000 years believe that. In general, the scientific attitude is to base one's beliefs on the evidence.

Questions have been raised about whether the condition of true justified belief is generally sufficient, as well as necessary for knowledge, and therefore whether it furnishes a definition. Circumstances can be imagined in which, as luck would have it, an agent can lay claim to true justified belief without, it seems, really knowing. Someone might truly have believed in 2007 that the last prime minister of Sweden had a surname beginning with "P", and justified this belief on the grounds that it follows from the belief held by that person that Sweden's last prime minister was Olof Palme. In fact, however, Sweden's last prime minister at the time was Göran Persson. Accordingly, the justified belief was true, but only by a fortuitous conjunction of circumstances, and not as the result of a cognitive achievement that having knowledge seems to be. Russell (1912, p. 76) drew attention to such cases, pointing out that for the purposes of defining what knowledge is, justified true belief is not knowledge when deduced from a false belief. It is not sufficient to require that the justification be a deduction from true premises, however; the premises must be known. But to require that such premises be not merely believed, as in the example, but actually known, would lead us in a circle in the attempt to define knowledge. Similar cases were presented by Gettier (1964), apparently in ignorance of Russell's contribution, and the general issue raised for an adequate definition of knowledge has come to be known as the Gettier problem. But even if incomplete as a definition, the condition of true justified belief certainly captures a substantial core of the notion, and that will be enough for present purposes.[5]

A further limitation on the above discussion is that it presupposes that knowledge is always someone's knowledge. But there are clear senses of knowledge which are not reducible to personal knowledge. We sometimes speak of what is collectively known to scientists in a certain field, which may not be something known in its entirety by any one worker in the field, but is recorded in accessible form in journals, books and data storage facilities and can be recovered by the experts should the need arise. Alternatively, the matter might be too complicated to be known by any one person, but is known to a team of people who complement one another's knowledge and can cooperate to manifest jointly knowing, for example in coming to grips with a problem. Some matter discovered at some point in time might have simply been forgotten by the scientific community in so far as no single person knows it at some later time. But should the relevant question arise, since a search of the literature would reveal the matter, we nevertheless say that it is known. Once more, knowledge in this sense is clearly related to evidence and justification which can be socially manipulated.

1.3 Inductive Support and Proof

There was a fundamental difference between the justifications of the two examples in the last section of the location of the sunspots and Fermat's last theorem, even if there were also some points of similarity. Although reasoning by more or less elaborate argument was a feature of both cases, Galileo's argument rested on his telescopic observations. There is no appeal to observation in the proof of Fermat's

---

[5]Some philosophers, notably Williamson (2000), take the view that knowledge can't be defined and must be accepted as a primitive notion. Not all terms can be defined on pain of going in a circle, and perhaps knowledge should be counted as one such indefinable, primitive notion, whose import is clarified by specifying its relation to other relevant notions. (Compare what is said about the notion of truth in Sect. 3.2.)

theorem. Galileo made a number of observations of the moving sunspots, from which he clearly concluded that their movements follow a general pattern and exhibit systematic changes. They always make their first appearance at the same edge of the solar disk, when they are thin, elongated bodies which fatten as they proceed towards the centre of the solar disk. He had no proof that they always behave like this—that one day, a sunspot might not behave differently and emerge from the opposite side of the solar disk, say. Mathematicians before Andrew Wiles had never found a non-zero integral solution to the equation $x^n + y^n = z^n$ for $n > 2$, but despite a growing number of worked examples, they never claimed to know that for all $n > 2$, the equation $x^n + y^n = z^n$ has no non-zero integral solutions. The mathematicians held out for a proof of their proposition before claiming justification. But there would have been no point for Galileo to hold out for a proof of his propositions. They were the sort of thing which simply can't be proved.

Popular writing about science sometimes suggests otherwise by speaking of "scientifically proven" facts. But this usage obscures the important feature of knowledge of propositions like those Galileo maintained about the sunspots. These are *empirical* claims, whose justification takes the form of mounting observational (empirical) *evidence*, which *supports* them but does not prove them. Mathematical claims are justified by *proving* them, which takes the form of *deducing* them from accepted axioms. Empirical results may be deduced from other empirical results, just as one mathematical theorem may be deduced from others. But ultimately, whereas mathematical theorems fall back on axioms and rules of deduction, empirical results are not deduced from the evidence but inferred on the grounds that it strongly supports them. Arguments in which the conclusion follows by deduction from the premises are called deductive arguments. Arguments in which the conclusion is supported by the evidence given in its premises are called *inductive arguments*.[6]

Very simple examples of inductive arguments are arguments by enumerative induction. These take the form of arguments in which the premises (the evidence) ascribe properties to particular individuals and the conclusion goes beyond this in some way, taking the form of a generalisation or a statement about what will be observed in the future. An inference from the fact that all swans observed up to a given point in time have been white to the conclusion that all swans are white provides a straightforward illustration. Notoriously, this inference, made before Western explorers discovered Australia, proved to be mistaken when black swans were observed there for the first time. Many inductive inferences drawn in the past

---

[6]The term is, unfortunately, ambiguous because there is a form of argument familiar in mathematics which is also called induction. An argument by mathematical induction, however, is a deductive argument, and not an argument by induction in the present sense. Since this is not the place to try to revolutionise standard terminology, suffice it to say that there will be no discussion of mathematical induction here, and "induction" as a form of argument will always be used in contrast to deduction. Needless to say, in physics the concept of magnetic induction (taken up in Sect. 2.2 below) is a further use of the term "induction" to be distinguished from that here contrasted with deduction.

have suffered a similar fate. Although investigations may seem to have covered all kinds of circumstances and the inferences then drawn were consistent with other inductive conclusions, future research revealed counterexamples and the inductive conclusion in question proved false. Here we can speak of being "proved" false because it can be deductively inferred from a counterinstance to a generalisation, such as the counterinstance "This swan is black" to the generalisation "All swans are white", that the generalisation is false and not all swans are white.

As Galileo's account of the sunspots amply illustrates, however, inductive argument is not confined to simple argument by enumerative induction. The overall argument is much more complicated, and even if it involves parts or subarguments which are simple enumerative inductions, there may also be parts which are deductive mathematical arguments. Since an argument is no stronger than its weakest link, however, the overall argument is not deductive, but inductive. It therefore carries the inductive risk that the conclusion may not hold in full generality. Just as in Galileo's case, we frequently do feel justified in accepting conclusions supported by evidence because we feel the evidence is sufficiently broad. But there is no escaping the element of inductive risk. We might endeavour to reduce the risk, but it cannot be eliminated altogether. There is always the possibility that the conclusion will prove wrong. This can't be circumvented by holding out for a deduction from the evidence, since deductions just can't be had for what are called empirical conclusions. The claims of science are not scientifically proved, but supported by abundant evidence.

Deductive arguments don't actually establish their conclusions beyond doubt. What they show is that the conclusions follow from the premises. This only shows that the conclusion is true if the premises are true. A similar point holds for inductive arguments. If it transpires that the premises—the evidence—are not true, then even if the argument shows that the premises would provide good grounds for the conclusion, the argument doesn't provide a reason for accepting the conclusion. In the case of empirical claims, we are thrown back on the evidence supporting them. In the case of mathematical claims, we are thrown back on the axioms from which the theorems follow. The Greeks thought axioms should be certain truths, which would be the case if they were simple, self-evident statements. Simplicity and the corresponding self-evidence of the fundamental axioms in mathematics cannot now be maintained as the Greeks thought it could. There is more on this theme in Sect. 3.3 below, and a response to this challenge is outlined in Sect. 6.4 of Chap. 6. Views on what is fundamentally a matter of empirical and what a matter of a priori justification have varied in the course of history and modern conceptions of empirical knowledge emerged from disputes about this especially during the sixteenth- and seventeenth-century scientific revolution. Some aspects of this development are mentioned in the next chapter. But for the moment, the important point is that empirical arguments are inductive, and carry with them the inductive risk that the evidential support, however good it is taken to be at one point in time, may later prove inadequate. This would still be the case, even if conclusions previously taken to be certain, such as mathematical truths, should turn out not to be so certain after all.

1.4 Looking Forward

The intention in this chapter has been to lay down a framework which will facilitate the elaboration and further discussion of scientific knowledge in the following chapters. Knowledge is true justified belief, where the belief is held for the reason that is given by the justification. Where the belief is an empirical one, motivated by appeal to evidence, the justification is inductive and does not ensure absolute certainty. It may, nevertheless, give grounds for great confidence.

Objectivity was briefly mentioned at the beginning of the chapter as something Galileo was concerned to establish, and this is an important feature associated with scientific knowledge. What does it amount to? Truth, as Galileo recognised, is one aspect of objectivity. Truth is independent of us, and the relativisation of knowledge to a knower comes via the mental attitude of holding true otherwise known as belief. But this aspect of objectivity has been questioned in recent years, inspired in many cases by historians and sociologists in the wake of Thomas Kuhn's *The Structure of Scientific Revolutions*. Such critics of the objective notion of truth are relativists, who think that the notion of truth should be relativised to a person or group of people. What is true for one person might not be for another. This line of thought is taken up in Chap. 3, where it is argued that, however appealingly liberal it may sound, it is fundamentally misguided.

A second aspect of objectivity concerns the justification for holding something true. Belief should be objectively motivated. My beliefs are mine, but if I claim to know them, then they must be objectively grounded in reasons others would accept. This has suggested to some writers that there should be as little contribution from the believer as possible, and the facts should be allowed to speak for themselves without being subjected to any kind of treatment by the believer. But the real world of science is more complicated than this would suggest, and simple formulas of this kind are misleading. This topic is taken up in the next chapter, where the justification of belief is discussed in much more detail. The chapter concludes by taking up some questions of how values should be respected in the development and reporting of science, and poses some questions about ultimately distinguishing facts and values.

The first part of the book is concluded with a chapter on the use and abuse of science, where moral issues are raised about the conduct of scientists and their responsibilities in the use to which the fruits of their labours are put.

The second part of the book looks at some philosophies of science. A contrast is drawn between those who see the progress of science primarily in terms of rejecting old hypotheses and theories and replacing them with new ones, and those who see science as progressing primarily by accumulating knowledge, saving as much as possible from older theories. The Austrian philosopher Karl Popper is a well-known representative of the first group, and the French physicist, historian and philosopher Pierre Duhem is a well-known representative of the second group. A concluding chapter discusses the natural attitude of taking the theories of modern science to be literally true, i.e. realism, in the light of arguments drawn from the history of scientific progress in favour and in criticism of this stance.

Chapter 2
Objectivity

2.1 Introduction

We think of scientific knowledge as objective, as meeting certain standards of objectivity, which involves staying in touch with the facts and guiding our thoughts by a concern for the truth. We strive after objectivity by not indulging in wishful thinking and allowing bias to distort our judgement. There is a paradigm of the disinterested observer, who records what is seen without allowing any personal evaluation of significance, appropriateness, and so on, to affect his judgement. We should, it seems, let the facts "speak for themselves".

The idea of a disinterested observer is illustrated by the cardiologist searching for indicators of coronary artery disease who noticed an association with earlobe creases (Frank 1973). He hadn't the faintest idea why the one should be associated with the other, but textbooks report the finding as a useful diagnostic sign, without any suggestion of a theoretical link between the two. In the same spirit of theoretical neutrality, the pharmaceutical industry screens new compounds, regardless of the purpose for which they were synthesised, for activity in a host of therapeutic tests. Most drugs have been discovered "empirically" in this fashion.

To take a historical example, Galileo discovered the moons of Jupiter with the aid of his telescope and was meticulous in recording their positions. As they revolved around Jupiter, he sketched pictures of his observations recording their positions in relation to the background stars which he carefully drew. Some four centuries later, historians examining his records note that one of the background stars is systematically recorded as changing position in just such a way as Neptune would have moved around 1610 when Galileo made his observations. Officially, Neptune was only discovered towards the middle of the nineteenth century. But there are some grounds for saying that Galileo saw and followed the movement of the planet, even if he didn't appreciate the full import of what it was he saw (Drake and Kowal 1980). It seems that Galileo is being praised as a disinterested observer who has the

© Springer Nature Switzerland AG 2020
P. Needham, *Getting to Know the World Scientifically*, Synthese Library 423,
https://doi.org/10.1007/978-3-030-40216-7_2

ability to notice matters of detail and report whatever observations have been made without any process of selection.

Even if due account must be taken of examples such as these, however, they are hardly typical or model cases of what the quest for objectivity involves. They misleadingly suggest that with minds sufficiently open to allow in whatever comes their way, scientific results will be caught as though by accident. More typically, the appropriate standards of objectivity are only met by interference in one form or another on the part of the investigator. Some commentators see in any contribution from the observer a retreat from the ideal of objectivity, and argue accordingly that the all-pervasive "interference" typical of normal scientific practice renders any claim to objective scientific knowledge preposterous. There are many ways in which observers fail to comply with the ideal of the passive bystander, and make substantial contributions to the observation process. Several are illustrated in the next section. So far from supporting this critique of objectivity, however, it is argued here that they show the critique to be based on a hopelessly naive, simple-minded vision of objectivity which there is no reason to take as a model for all knowledge. Objectivity shouldn't be confused with hygiene. Standards of objectivity are not a matter of certifying that goods are untouched by human hands. The concern is not the avoidance of interference, but of adequately dealing with it. This requires the ability to critically examine the circumstances, appreciate the relevant factors and take them into account. As the consequences are drawn out in subsequent sections, a social dimension of objectivity emerges—the need to justify a claim to knowledge involves arguing the case before a critical public.

2.2   Observation and Experiment

Let us begin with a simple case in which the tendency to bias on the part of observers is confronted with an active response to meet the problem. At the beginning of the twentieth century, several kinds of radiation had been newly discovered. These were called α-radiation, β-radiation and γ-radiation, each with distinct features which subsequent research has penetrated and elucidated. In 1903, the work of a team under the direction of René Blondlot, professor of physics at the university of Nancy in the east of France, came to the attention of the scientific community at large. They reported the discovery of a new kind of radiation which they dubbed "N-radiation" in honour to their university town. But interest dwindled as other investigators found it impossible to reproduce their results. In the end, a Canadian spectroscopist, R. W. Wood, travelled to Nancy to follow the work of experimentalists there with his own eyes. The effect that the Nancy observers set great store by was the enhanced brightness of an electric spark, which the they took to confirm that the discharge occurred in the presence of the N-rays. But try as he might, Woods could not see the enhancement which the Nancy observers claimed to see. His remedy was to suggest that the observers report on the brightness of the spark under circumstances in which they are in ignorance of whether the electrical discharge is exposed to the supposed radiation source. A suitably randomised series of experiments was conducted in which the electric spark was sometimes exposed to the supposed source of radiation, sometimes not, and the observations of the spark subsequently compared with the exposure data. It was thus possible to check the reported brightenings against the cases when the spark was actually exposed to the supposed source of radiation, and see whether there was a positive correlation. The conclusion of Wood's investigation was negative (Wood 1904; Klotz 1980).

Repeatable phenomena, subject to general laws or principles, are often the subject of scientific investigation. Where this is so, reproducibility in experimental work is a necessary, but as we will see, not a sufficient, criterion for a justifiable claim to knowledge. Some observations are difficult to perform, and require a training which may itself explain why reproducibility is difficult. But in the N-ray case, careful study of the original observers revealed a source of bias which could be eliminated, once appreciated, by a procedure of statistical randomisation. Unfortunately for Blondlot, striving for objectivity in this way undermined the justification of his claim.

In general, we have a formula: recognise and accommodate! But this is not much of a practical guide because there are so many ways in which the observer fails to maintain a neutral attitude and makes an active contribution to the production of scientific knowledge. It is not the purpose here to offer an exhaustive classification, but grouping illustrative cases under a few headings will serve to introduce the sort of considerations involved. Accordingly, we might distinguish cases in which psychological factors enter into the observer's perception, cases in which the observer's expectations guide the investigation, and, finally, cases in which systematic sources of error are recognised and integrated into the results. These will be considered in turn.

Galileo realised that science cannot be based simply on what, as he put it, is "evident directly to our sight, without any need of reasoning" (Galileo 1613, p. 107). He came upon this problem time and again, realising that those unfamiliar with the trials and tribulations of conducting careful observations were naturally suspicious of appeals to "doctored" results as evidence which should persuade them of his newfangled ideas. He had to engage in a debate with recognised authorities to argue his case, and published his results in the form of dialogues and correspondence in which he openly presented and tried to meet objections. One such publication is the *Letters on Sunspots* (1613), where he sets out for public scrutiny a correspondence originally conducted via a third party with an opponent who used the pseudonym "Apelles" (later identified by historians as the Jesuit priest Christopher Scheiner—Drake 1978, pp. 210, 233, 267–268). One topic concerned the appearance of sunspots. Surely they are black. This was not Galileo's view, however.

Galileo argued that an illusion of darkness is created by the sharp contrast with the immediate background which must be corrected. He points to a number of comparisons, arguing that since sunspots are no darker than the region immediately surrounding the sun, which is in turn not darker than the moon, which is certainly not darker, but brighter, than the darker regions on its surface, then the sunspots are, contrary to Apelles' view, brighter than the darker regions of the moon. What is remarkable in this case is that even the appearance of the sunspots on which Galileo based his inferences is not settled without argument. For surely Galileo was right to challenge Apelles' understanding of the simple appearances and submit even those to scrutiny. It is now a standard result in the psychology of perception that awareness of colours and shades of lightness and darkness is affected by the immediate surroundings of the object in question. But this was not generally recognised at the time, and Galileo's own reflection led to his discovering this "context dependence" of observation and how to take it into account.

In other cases, Galileo found that the appeal to observation could be even less straightforward, especially when it involved the idea of formulating terrestrial physics in mathematical terms, which, if not entirely revolutionary, was still controversial. Although he felt the degree of agreement he managed to achieve was "superior", as Naylor (1974, p. 41) says, "to anything obtained previously", it would have been asking too much to expect his opponents to cock a sympathetic ear to all the difficulties of conducting precise experiments. He called for precise experiments to distinguish competing theories, but many sources of error arise which make observation a far more complex affair than was recognised by the naive Aristotelian conception of appeal to sensory experience.[1] This raised the problem of justification. Why should his opponents discount observations conflicting with his theoretical claims and accept Galileo's intricate explanations of these "anomalies" if support is to be derived from observation? Didn't Galileo's attempts to explain away recalcitrant observations display the same unwillingness to face up to the evidence of our senses as did those Aristotelians who refused to look at the sky through his telescope?

A second example of the way psychological factors influence observations, from a later period when the need for accommodating relevant factors was more generally recognised, is provided by the case of the dismissal in 1796 by Maskelyne, the observer royal at the Greenwich observatory, of his assistant Kinnebrook. Kinnebrook's observations of stellar transits differed by almost a second from Maskelyne's own. After Maskelyne warned Kinnebrook for recording times a half-second later than his own, Kinnebrook's striving to correct his erroneous technique only increased the discrepancy and he was dismissed. The error was significant because the calibration of clocks would depend on it in Maskelyne's suggested method of determining longitude at sea (Sobel 1995). The standard procedure for observing stellar transits was Bradley's "eye and ear" method, known to reproduce the results of a single observer to within a tenth of a second. The field of the telescope was divided by parallel cross-wires, and the observer glanced at the clock,

> noting the time to a second, began counting seconds with the heard beats of the clock, watched the star cross the field of the telescope, noted and fixed in his mind its position at the beat of the clock just before it came to the critical wire, noted its position at the next beat after it had crossed the wire, estimated the place of the wire between the two positions

---

[1] See Naylor's comments on Galileo's pendulum experiments in Sect. 4.1.

> in tenths of the total distance between the positions, and added these tenths of a second to the time in seconds that he had counted for the beat before the wire was reached. It is obviously a complex judgment. Not only does it involve a coordination between the eye and the ear, but it requires a spatial judgment dependent upon a fixed position (the wire), an actual but instantaneous position of a moving object, and a remembered position no longer actual. (Boring 1957, p. 135)

The event was recorded in *Astronomical Observations at Greenwich*, and came to the attention of Bessel some years later when reported in a German history of the Greenwich observatory published in 1816. A new observatory had been built at Königsberg in 1813 under Bessel's supervision, and he suspected that Bradley's method might be subject to an idiosyncratic personal error, which would not affect the reproducibility for a single observer. He investigated whether similar discrepancies could be found amongst his own colleagues, and discovered that they varied idiosyncratically. Comparing himself with Walbeck when observing stellar transits for ten selected stars, for example, he found he consistently recorded earlier times than Walbeck, the difference varying little from the average of 1.041 s. Bessel devised a method of calibrating observers by reference to a third observer issuing in a "personal equation" with a view to correcting for the observer's idiosyncrasy (Boring 1957, pp. 137–8), and the differences have since been much debated by psychologists. Like the appearance of sunspots, we have here an involuntary source of influence on observations which an astute scientist became aware of and tried to accommodate in determining the results of observation. Modern techniques strive to replace human observers with recording machines not subject to such influence.

Turning now to the second group of cases, expectations on the part of investigators play an important part in determining what counts as a genuine result with an accepted role in a body of scientific evidence. Consider the sad case of Jean Daniel Collodon, who failed in 1825 to do what Michael Faraday to great acclaim succeeded in doing in 1831, discover magnetic induction. In 1820 Ørsted had shown that an electric current affects a magnet, presumably by giving rise to a magnetic field. This led to the development of the galvanometer which measures the strength of an electric current by its ability to turn a magnet and twist the wire by which it is suspended. Physicists suspected that there would be a converse effect in which a magnet gives rise to an electric current. They expected a relatively long-term effect, however, which blinded them to the significance of some of the things they actually saw and recorded, and played an important part in how they devised their experiments and conducted their observations in pursuit of the effect. Collodon considered first an apparatus comprising a wire coil attached to a galvanometer in a circuit, allowing the registration of any current appearing when a magnet is brought close to the coil. But he was afraid that this magnet might affect the magnet in the galvanometer directly, so that any reading wouldn't be unambiguous evidence for the existence of a current in the circuit since it might be caused by purely magnetic interaction. Accordingly, Collodon took the precaution of extending the circuit, and connecting the coil to a galvanometer placed at some distance in an adjoining room. Satisfied with his set-up, he proceeded to make his observations, carefully moving the magnet up to the coil, and then going into the next room to see whether the galvanometer was recording any current. Needless to say, he saw no signs of a current.

Faraday, on the other hand, had different expectations, based on his conception of a magnetic field as consisting of lines of force emanating from one pole, curving back in the space around the magnet, and ending at the other pole. He thought that what was needed to give rise to an electric current was the breaking of these lines by the coil as the magnet is moved in relation to the coil, whereas a stationary state of the lines interlacing the coil would not cause a current in the circuit. So he expected a short-lived effect, and prepared himself accordingly. Had Collodon thought in the same way, he could easily have arranged for an assistant to move the magnet while he observed the galvanometer himself and preempted Faraday's discovery. Historians examining laboratory notebooks of Collodon's and Faraday's contemporaries have noted that short-term "disturbances" of galvanometers were in fact seen by investigators who dismissed these effects as unsystematic influences which more careful experimentation would eliminate.

With greater confidence in their knowledge of underlying causes, scientists have in other cases been more successful in making discoveries by following up unusual phenomena. One sunny day early in the second decade of the twentieth century, Harry Brearley was enjoying a stroll in his native Sheffield when he came upon a rusty heap of scrap metal and a bright reflection caught his eye. He was intrigued, because as a metallurgist he realised what many others, who must have walked by and seen what he had, evidently did not, that metal cheap enough to be found on a scrap heap would be tarnished and rusty after exposure to the weather. He retrieved the shiny metal fragments and analysed them, finding that they comprised a mixture of chromium and iron. Understanding that the chromium had prevented the iron from rusting, he experimented to find the optimal proportions, and so discovered stainless steel. (It transpired that the original scrap metal derived from gun manufacturers, who were experimenting on a purely trial-and-error basis with artillery made from various metal mixtures in a search for stronger material suitable for larger guns.)

We can take up the third class of cases with the consideration that Collodon's precautions prevented him from discovering magnetic induction given that he was working alone. But his general strategy of considering the various ways in which his apparatus could be influenced was sound, leading him to take precautions in an attempt to ensure that it responded uniformly to the phenomenon he was investigating in ways he understood. Reproducibility is the acid test of reliable data. And if other investigators are to be able to confirm results reported by a given group, then it must be possible to set up the experiment or carry out the observations in the same way. Wayward influences must be eliminated by procedures which are open to evaluation, and effects which can't be conveniently eliminated must be systematically taken into account in the interpretation of the results in ways which are open to criticism and don't raise suspicions of cooking or trimming the data. As Pierre Duhem (1861–1916) points out in discussing Regnault's mid nineteenth-century work to improve on Boyle's law for the compressibility of gases,

> ... when we criticize Regnault for not having taken ... [the action of weight of the gas under pressure] into account ... [w]e are criticizing him for having oversimplified the theoretical picture of these facts by representing the gas under pressure as a homogeneous

fluid, whereas by regarding it as a fluid whose pressure varies with the height according to a certain law, he would have obtained a new abstract picture, more complicated than the first but a more faithful reproduction of the truth. (Duhem 1906b, p. 239; Eng. trans., p. 158)

Another example of such criticism came after Albert Michelson published the negative results of his first experiment to determine the earth's velocity through the ether in 1881. H. A. Lorentz pointed out that Michelson had omitted the non-zero effect of the ether wind on the transverse of the two perpendicular arms of his apparatus. Appreciating that moving across a current and back takes longer than crossing over still water was one of the factors which led Michelson to design and build a better apparatus when he later collaborated with Edward Morely.

There is no question of phenomena being "evident directly to our sight, without any need of reasoning", as Galileo put it. This was not possible, as we have seen, even in Galileo's time, and empirical investigation has certainly not become simpler since then. These points can be illustrated by considering one of the most famous experiments in the history of science, Millikan's experiment on oil drops to measure the charge on the electron.

The electron was discovered in 1897, at a time when physicists were making great strides in the care with which they conducted experiments and increasing the precision of their results. Millikan's problem was to devise a way of producing bodies with a small charge which he could measure. If the charge on each body was due to a collection of not too many electrons, it might be possible to find a common denominator, and infer the charge on a single electron. Alternatively, observation of small changes in charge might similarly reveal a common denominator. The friction generated in producing small oil drops gave them a negative static charge which he could measure by allowing the drops to fall in an electric field between two plates. The upper plate was positively charged, retarding the rate at which the drops fell under the influence of gravity. By adjusting the field between the plates, he could measure the strength of the electric field corresponding to a given measured terminal velocity of fall, which is related to the net force acting on the drop. The only unknown factor in this net force is the charge on the drop, since the diameter of each drop could be measured, which, assuming the drops to be spherical, gave the volume, and then, given the density of the oil, the mass was forthcoming. The downward force due to gravity could thus be calculated, from which the upward electric force could be subtracted to yield the net force on the drop. It seems, then, that the charge on the drop can be calculated from the equation relating the terminal velocity to the net force on the drop. Performing a series of experiments enabled the charges on different drops to be compared and a common denominator found which was taken to be the charge on a single electron.

But Millikan thought that the situation was considerably more complicated than this basic picture would suggest. For one thing, there were more forces at work. The small oil drops were very light, and easily disturbed by any movement of the air. Millikan enclosed the whole apparatus in a sealed container to reduce air movement. But he still had to contend with convection currents caused by uneven temperature, which he reduced by maintaining a constant room temperature with a fan. The choice of oil was the result of previous trials initially with alcohol, which proved unsuitable because of evaporation and consequent variation in mass of drops in the course of measurement. As a result of much preliminary work of this kind, he was eventually able to modify his apparatus and procedures so that he could obtain consistent, reproducible results. But his troubles didn't end there.

Reproducibility is not a sufficient criterion that no further forces are in operation affecting the course of the experiment. There might well be systematic effects which, unlike air turbulence, consistently act in the same way, and shift the final calculated result from the true value. In the best of all possible worlds, several such effects would cancel one another out. But it would be wishful thinking to suggest that all such effects can be safely ignored for this reason. Clearly, they should be either eliminated or explicitly taken into account in the calculations. Amongst such effects is the Archimedian up thrust experienced by any body in a fluid medium proportional to the amount of fluid displaced. A second effect, more difficult to quantify, is the resistance to movement offered by the air's viscosity. Millikan was dissatisfied with Stoke's classical theory, which he replaced with a modified empirical law providing a more precise equation for the circumstances of his experiment. Both the viscosity of air and the density of the oil varied with temperature—in a systematic fashion which could also be taken into account in the calculations. But it introduced the need to measure the temperature within the closed container (Franklin 1986, pp. 215–25).

Millikan wasn't able to completely eliminate spurious factors giving rise to a small variation in successive results calculated from repeated measurements. But having satisfied himself that his apparatus was working within a reasonably small range of variation in accordance with principles he understood and could take into account, he was prepared to begin taking measurements in earnest. The small variation in successive calculations would have to be accommodated statistically, taking an appropriate average of all his results falling within an estimated range of possible error. This was incorporated into an overall estimate of *precision*, taking into account the ranges of possible error arising from each of the factors contributing to the final calculation. His final result was reported in the form of an approximate judgement, that the charge on the electron falls somewhere within a certain range, delimited by the margins of error: $(4.774 \pm 0.009) \times 10^{-10}$ esu.

The claim is not simply that the value of the electronic charge lies within the estimated range of error. The "$\pm$" error is a statistical claim that the probability of the value lying within that range is such-and-such. For a given experiment, a higher probability is achieved by broadening the range whereas narrowing the range lowers probability, and standard conventions for the probability are usually adopted (see Hughes and Hase 2010 for more details). Improved techniques that narrow the range of error for the same probability of the value lying within it increase the precision of the experiment. Duhem admired the increased precision of Boyle's law achieved by Regnault's improvements on the experimental technique. His critique was directed towards the *accuracy* of Regnault's results due to overlooking a systematic error. The range of error determined by the *precision* of an experiment is centred on the reported value and systematic errors, which shift this value, are said to affect the *accuracy* of the result. This distinction between accuracy and precision allows that a very precise experiment (narrow range of probable error) might nevertheless be inaccurate, or a more precise experiment might be less accurate. Alternatively, an imprecise experiment (broader range of probable error) might nevertheless be quite accurate.

It might be thought that results reported in this fashion are vague, not committing the experimentalist to an exact value. But to do so would be to appeal to a certain vision or ideal of the standards of articulation of what we can claim to know which doesn't belong to this world. We will see shortly that the possibility of objective knowledge is easily criticised from the vantage point of unrealistic standards which can't possibly be achieved. More interesting is a critique of a concept of knowledge which does have application. In the present case, we can say that times have definitely changed since the days of Galileo, and the role of systematic and random error is now accepted as a matter of course. Far from being vague, it is a mark of precision that the results can be specified with a well-articulated margin of error. Results reported with no margin of error, and no indication of the likely circumstances which would invalidate the results and how these have been accommodated—either by eliminating them or incorporating them into the uncertainty of the result—are, on the contrary, vague. Their value to the scientifically literate reader is unclear, and give the impression that the investigator doesn't really understand what he is doing. In any case, the idea that there is a definite value to, say, the length of a rod in centimetres as literally specified by a real number is strictly speaking meaningless. When we leave the macroscopic realm and get down to the microlevel, molecules don't occupy definite regions because the electrons surrounding the central structure of nuclei are not entirely confined within any given boundary. Chemists typically think that the distribution of 95% of the electron density gives a pretty good idea of molecular shape. But there is no spatial boundary beyond which the probability of finding an electron falls to zero.

Even where there is a quantitative, statistical estimate of the random error, there will also be a qualitative aspect to this formulation of error, covering the various contributions to systematic error. The qualitative aspect remains where, as in the case of scholarly work, measurements and statistical spreads are not at issue. Unusual aspects may be explicitly discussed in published reports, but much is tacitly understood by the scientific community and doesn't appear in the journals. For reasons of space, arguments are often suppressed in modern scientific publications. But many considerations are standard, and familiarity with them forms part of every scientists' training. Still, all aspects are open to review.

2.3 Objectivity of Interpretation

Interpretation is sometimes thought to be the sole prerogative of the humanities. Understanding the phenomena which constitute a culture may call upon many special features, and so incorporate special features into interpretation. But the general notion as such has wider application. Writing at the beginning of the twentieth century, the physicist and historian of science Pierre Duhem summarised his review of considerations of the kind taken up in the last section in a philosophical work, where he describes an experiment as

> the precise observation of phenomena accompanied by an interpretation of these phenomena; this interpretation substitutes for the concrete data really gathered by observation abstract and symbolic representations which correspond to them by virtue of the theories admitted by the observer. (1906b, pp. 221–2; Eng. trans., p. 147)

Raw observations don't stand on their own feet, but become the considered result of empirical investigation only after interpretation in the light of the experimenter's understanding of the phenomena.

The comparison shouldn't be taken to absurdity, and the special features just mentioned observed. There is no question of the experimenter arriving at a view of what the instruments are thinking. Consequently, there is no place in the interpretation for the role of assumptions of rationality involved in the understanding of an action. Where human action is at issue, the interpreter explains the subject's actions as directed towards some goal. This involves the ascription to the agent of reasonable beliefs and desires, with a view to showing how the action could be seen as a reasonable thing to do from the subject's perspective. The English philosopher Collingwood described this by saying that an action's "outside"—the observable behaviour—is explained by giving an account of the action's "inside". Although providing a description of the subject's view is not to be confused with subjectivism on the part of the interpreter, questions might nevertheless be raised about the objective status of the interpreter's interpretation.

Traditional ways of speaking about the interpretation of actions certainly do raise worries about objectivity. It was suggested that the interpreter must abandon his own view of the world and "put himself into the agent's shoes" by some process of mental projection, enabling him to "feel" exactly how the agent himself did in the foreign or historic circumstances in which the action was performed. With no reference to external criteria, this process of mentally reliving the life of the agent provides for no checks and balances, and does not allow a distinction, even in theory, between the interpreter's view of the agent and a correct view of the agent. Even as the interpreter's own view of the agent develops and changes, all we can say is that one view is later, and perhaps more elaborate, than the other; there is no principle by which the one can be counted more adequate than the other, except in so far as the interpreter himself says so.

This general line of criticism inspired the behaviourists in the first half of the twentieth century to try to do away with descriptions of mental phenomena altogether. There is no support for the ascription of any mental feature, they argued, independent of the phenomena it is devised to explain. This implies that such ascriptions have no independent meaning; they have the sense of "the cause—whatever it might be—of the behaviour in question", so all the mental ascriptions tell us is that behaviour is caused by whatever it is that causes it. The behaviourists sought to remedy this evil by abandoning the mental as an inscrutable black box, and concentrate instead on systematic correlation between stimulus and response—the observable physical input and the observable human behaviour prompted by the stimuli. Intense research enriched our wealth of knowledge with an insight into the capacities of rats and doves to turn left or right when confronted by a red light set against a plain white background (anything more interesting would throw into doubt what the actual stimulus was).

In a famous critique of the application of these simple-minded ideas to the elucidation of one of the most characteristic and fundamental features of human behaviour, the ability to command a language, the American linguist Noam Chomsky (1959) pointed out that the behaviourist strategy is impaled on the same argument that was used against the use of mental descriptions. There is, in general, no way of identifying which stimulus, from all the phenomena, past and present, that impinge on the subject, actually prompted a given verbal response without additional help, usually in the form of an understanding of the verbal response itself! A laboratory situation in which there is only one feature which is plausibly taken to be the stimulus is not the situation in which humans act. In real life, an agent is currently affected by many influences and aware of his own considerable history. Which stimulus actually prompts an agent into action will depend, among other things, on what he actually notices. But in order to sort out what is noticed from among myriad potential influences bombarding the agent, it is necessary to resort to the agent's perspective, and consider what captures his attention. Simplified circumstances might be envisaged, appropriate to what a field linguist might be expected to take as his starting point in learning a radically new language from scratch, with no cultural history of translation to fall back on. But this would, at best, provide translations of only a few whole sentences amenable to correlation with observable phenomena (Quine 1960). Even then, the responses would be interpreted on the assumption that what the field linguist feels reasonably sure the agent has noticed are guided by appropriate beliefs and desires. Suppose the agent believes what the linguist believes is clearly visible and salient; still he might not understand the linguist's project of trying to master his language, and even if he does, he might desire to mislead or play games with the linguist. (Children in Iraq amused themselves after the invasion in 2003 by approaching and swearing at American soldiers, who misinterpreted the children's words more favourably.) The correlation of stimulus and response in the absence of any consideration of mental state is therefore not going to provide a replacement for explanations of action in terms of motives.

But even if behaviourism has not provided a satisfactory alternative, the interpreter's project must be understood to be not susceptible to the objection of subjectivism. Broadly speaking, this is achieved by recognising that the interpreter's account of the agent's perspective is, like the experimenter's account of his experiment, a distinctly third person account. It doesn't seek to reproduce the agent's own feelings on the basis of some mysterious process of empathetic introspection enabling the interpreter to enter the agent's own mind. In the extreme case, it might not even agree with the agent's own account of the matter. A person who leaves home with an umbrella is naturally interpreted as believing that it might rain. If the question was put to him and he were to deny having this belief, some explanation of the extraordinary reply would be called for. But if there is no independent reason for thinking him mad, or unable to judge the state of the weather, or simply not co-operating with the questioner, then we must seriously consider whether he knows his own mind. An agent who does not want to admit an unpleasant truth, but nevertheless acts on the basis of certain beliefs and desires, is reasonably ascribed the beliefs and desires which best explain the action.

Beliefs must form part of an intelligible and coherent overall picture, which must be integrated with beliefs normally held in the agent's social and cultural sphere, and with beliefs forming part of the explanation of the actions constituting the agent's life. This holistic view of belief offers many points of reference against which any particular ascription of belief can be checked. It is on such considerations that the interpreter must base his account and recapitulate when called upon to justify his explanation to critical colleagues.

Empathy enters, not in the form of a special kind of introspection, but, for example, in the form of the adoption of a principle of charity—the view that an interpretation which portrays the agent as irrational, especially with respect to relativity simple matters, is better taken as a reflection on the poor quality of the interpretation. This rules out the possibility of discovering that inhabitants of a strange culture are so stupid as to contradict themselves. Anthropologists of bygone times have been known to make such claims; but a misunderstanding on the part of the interpreter of the construction of negatives in the foreign language, for example, or of some of the foreigner's general beliefs about the workings of the world, is more likely. This imposes our logic on the agent—the logic which the interpreter can display to his colleagues. But there is no alternative. Claims to have made empirical discoveries to the contrary fall on the grounds that there is no independent way of establishing the correct interpretation of negation. So far from abandoning his own world view, the interpreter must presume that the agent is likely to share the same beliefs about everyday matters, to be motivated by similar everyday desires and to reason in accordance with the same general logical principles. These are not immutable principles; but breaking them requires special explanation—relating, for example, to the fact that the agent lived in a different epoch (perhaps using a common word differently), or lives in a different culture. They bear on the question of how the agent can be rendered understandable to us. The interpreter must be able to provide an account in his own language, on the basis of his own concepts and beliefs, of any points of difference with his subject as best he can. The upshot may be a broadening or a changing of the interpreter's own perspective, perhaps by incorporating a foreign word or phrase into his own description, or as the result of a successful challenge to some of his previous views. Even so, it must be articulated. None of this is possible without a massive fund of shared belief.

2.4 Fallibility

The quest for greater precision leads to a search for sources of error which can never be definitively concluded, and the efforts of one group will inevitably be open to criticism from others. We have seen how Regnault and Michelson were taken to task for omitting to take into account factors which could be accommodated in subsequent investigations once they had been brought to light. Can the interpreter of an ancient text criticise the author on the basis of an anacoluthon, or is this evidence of a corruption in the text introduced by later transcribers? Does the impossibility of giving an absolutely conclusive argument, guaranteed immune from any future criticism, mean the scientist is never in a position to claim to have gained knowledge—that investigators can only report their opinions?

An investigator must be reasonably confident before making any claim to new knowledge. But confidence is a matter of settling reasonable doubt. There is no question of certainty. Given previous history, who would expect his claims to stand the test of time and not require revision or outright rejection in the long run? But although past experience suggests that it is a fair bet that future modifications will be required, this is no objection to offering presently justified claims as claims to knowledge.

The ancients thought knowledge is only possible of what is necessarily true. Philosopher-scientists involved at the beginning of the seventeenth century scientific revolution thought we could only really know what is certain and beyond all question of doubt. But these venerable attitudes were eventually dropped by those who have made it their business to think seriously and systematically about epistemology. And the brief examination of the critical attitude towards observation in modern science in the previous sections serves to confirm this picture of the fallibility of knowledge based on observation. Claims to knowledge are not threatened by once popular arguments to show that they don't meet impossible and unreasonable standards of certainty. The scientist who is aware of what he is doing shouldn't flinch from claiming, as a result of his investigations, to know something. Of course, he may not have satisfied himself beyond reasonable doubt that he does really know. Perhaps a suitably guarded suspicion, or an interesting hypothesis, is the best he can justifiably muster. But where he is confident in the reliability of his investigation, he can claim to know. The possibility of critique requiring revision is not grounds for agnosticism, but for an undogmatic attitude in recognition of the fallibility of knowledge. Though he claims to know, the investigator usually can't claim to know that he knows—to know that his claim to knowledge is beyond criticism. Confidence is one thing—a willingness to defend a thesis in the face of criticism so long as criticism can be reasonably rebuffed, without dogmatically holding onto a claim in spite of the criticism, leaving reasonable objections unanswered. Certainty based on knowledge that there couldn't possibly be grounds for criticism is quite another thing.

2.5 In the Eyes of Others

Another way of unreasonably questioning the possibility of knowledge from standards impossible to meet is to raise the spectre of an absurd vision of objectivity. How can the subject even know whether what he envisages—what he claims to know—is really true? For, so the objection runs, the subject is not in a position to step out of his own mind, as it were, and adopt the perspective required to make the comparison. There is no need to countenance the prospect of a divine overseer in order to envisage the appropriate "third person" perspective, however. The issue doesn't hang on a question of theology. We need look no further than a third person in the flesh, or preferably several. The locution of claiming to know expresses the idea that having knowledge puts the subject in a position to make a claim—to face the public and say I know. Knowing is a matter of being able to articulate a sentence the truth of which third parties can grasp and check themselves. As Kant put it, an objective judgement must have a content that "may be presupposed to be valid for all men" (*Prolegomena*, section 19; Kant 1783 [1966]).

This is obviously appropriate where a reasonable requirement of objectivity is reproducibility. But in many cases, the matter at issue is not a repeatable state of affairs. This doesn't mean that Kant's dictum ceases to be applicable. There may be the possibility of finding independent witnesses of a single occurrence. The streets of Uppsala filled with howling students after the outbreak of the French revolution in February, 1848 in a manifestation that worried the authorities. This could hardly have passed by unnoticed, and several newspapers gave (widely differing) accounts of the event. Perhaps there are no independent witnesses and the third party may have to fall back on the reliability of the subject as sole witness. Here the past record of the subject, and any particular circumstances which could affect the subject's judgement, or motives which may dispose him to lie, must be taken into account. The subject himself can also step back and assess his own judgement in the light of such considerations.

Recognising that claiming to know is something that is, in principle, done in public and under the scrutiny of others may help us avoid thinking that knowledge must be certain. But this may seem only to take us out of the frying pan into the fire. For doesn't the recognition that there are others whose views we must respect when trying to win their agreement by presenting the evidence which we think justifies our claim to knowledge raise another problem for the conception of objective knowledge? There is considerable variation in human belief. But how can we account for this diversity if the nature humans confront is the same? Shouldn't the same world give rise to the same beliefs? Yet surely the idea that we—several of us—can claim to have objective knowledge implies that our several claims to knowledge all concern the same world. Then it seems that the commitment to take seriously the views of others in public debate creates a dilemma given the fact of variation of belief. If we are committed to taking the views of others seriously, can't we best explain this diversity by giving up the ideal of objective knowledge?

It is far from clear that the appropriate view is to throw objectivity overboard merely because of disagreement. Consider the following example. For many centuries, vulgar opinion had it that the earth is flat. This is the natural result of simple observation of the observer's immediate environment. We can imagine how people of this opinion tried to dissuade Magellan from embarking on his voyage beyond the known world in view of what they feared might happen when he reached the edge of the earth. But this belief had been countered by another since antiquity. We know, for example, how Eratosthenes (276–194BC) measured the earth's circumference in the third century BC. The earliest surviving references to measurements of the earth's surface appear in Aristotle's writings, dating from a century earlier. Aristotle argued in the fourth century BC that the earth must be round. He pointed out that during an eclipse, the earth casts a round shadow on the moon. Further, the height of stars above the horizon (at a given time of the night) varies as the observer moves North or South. Finally, the upper parts of ships approaching port are seen above the horizon before the hull (*De Caelo*, II.13).

Neither instruments nor a knowledge of mathematics is required for the simple qualitative claim that the earth is round. So the argument for a round earth doesn't depend on matters not available to the layman. Perhaps the flat-earthers simply didn't think of looking towards the heavens to discover something about the earth. But it seems they didn't pay such careful attention to events on the horizon as did the round-earthers. Aristotle's appeal to the height of stars above the horizon and his appeal to the sightings of the upper parts of ships on the horizon are each a matter of considering a set or sequence of observations rather than making an inference from what is seen in a single glance. And in bringing the three areas of observation together in support of the same conclusion Aristotle went further and considered the combined import of several classes of observation. The situation emphasised in the present objection to objective knowledge, in which several people manifest diverging opinions about the world, can arise in much the same fashion for a single observer's several observations. But Aristotle clearly took it that all his observations concern the same world, and so they cannot stand in conflict with one another. But they cannot easily be reconciled with a belief in a flat earth. Those with neither the leisure nor inclination to take account of the several observations Aristotle reported would not be troubled by any such conflict. But intellectual curiosity in Aristotle's time led his contemporaries to note such phenomena, and Aristotle clearly thought that he had to make systematic sense of all these observations. The best way to do this was to understand them as phenomena arising from the same underlying fact that the earth is round.

On the basis of this belief, it seemed reasonable to set about measuring the size of the earth by measuring the angle light rays from the sun make with a vertical stick at Alexandria when the sun is directly overhead at Syene, a city 5,000 stades due south of Alexandria. Since the angle Eratosthenes is reported to have measured is 1/50th of a full circle, then on the assumption that all light rays reaching the earth from the distant sun are parallel, the measured angle is equal to the angle subtended by the arc stretching from Alexandria to Syene at the centre of the earth. Accordingly, the distance between Alexandria and Syene is 1/50th of the circumference of the earth, which is therefore 250,000 stades. If we knew how long the Greek unit of a stade is (which can't be easily deduced because the reported measurements have obviously been rounded off to make the report easier to write and read), then we would know what the Greeks claimed the circumference of the earth is in the third century BC. (Contemporary scholars think this to be about 5% less than the modern estimate.)

The return of what was left of Magellan's expedition in 1522 from the opposite direction to that in which it set out may have converted many flat-earthers. But for some two millennia people had differed on the shape of the earth. Should any of the ancient Greek intellectuals who agreed with Aristotle, or the Arab commentators who preserved the Greek literature throughout most of the middle ages, have accepted that belief in a flat earth was on a par with their own and refrained from claiming to know that the earth is round? Of course not. The flat-earthers simply didn't address the issues which led the round-earthers to their view. On the other hand, the round-earthers could explain the appearance, after a single unreflective glance, of a flat earth in terms of the relatively large size of the earth in relation to the observer. In general, people form beliefs on the basis of what they observe and what they infer from their experiences. And different people may well differ in their range of experience and their powers of reasoning. Such differences are often sufficient to explain differences of opinion, and circumvent the need to question the ideal of objectivity. So the idea that objective knowledge requires articulating the claim in a form that a third party can grasp and assess is not simply a matter of convincing any Tom, Dick or Harry. Rather than abandoning our claim to knowledge in the face of a dissenting party, we may cease to respect the dissenting party as a reasonable assessor of our claim. This doesn't, of course, imply lack of respect in anything other than an epistemic sense. Naturally, such conclusions don't come easily. It may well require some considerable time and effort to arrive at an explanation of our opponent's dissenting view in terms of different experience or powers of reasoning. But there is no dogmatism in coming to the conclusion that one of a pair of opposing views is true and the other false. Perhaps our opponent can convince us to retract our view; perhaps we don't think the criticism holds water. And it may be that in some cases we can't arrive at any definite conclusion, recognising that there is something in the opponent's view even if we remain to be convinced that our own is wrong. Here the undogmatic scientist should suspend judgement.

Much the same considerations can be put diachronically. Beliefs change; i.e. people hold different beliefs over time. Sometimes the same person comes to hold a different opinion with the passage of time—in the light of new evidence or further reflection or belated appreciation of another's view. Diachronic differences have been seen as a source of further questioning the notion of objectivity, as we will see in the discussion of relativism in Chap. 3.

2.6   On the Shoulders of Others

In contrast to the situation prior to Magellan's voyage, vulgar opinion today has it that the earth is round. Does this mean that ordinary people have examined the shadow cast by the earth on the moon, pondered the change in elevation of stars with the observer's position in a north-south direction and carefully observed the approach of tall ships in arriving at their belief about the shape of the earth? Hardly. They've been told that the earth is round at school, heard it on the radio, read it in the newspapers and comfortably acquiesced in agreement with friends and acquaintances. Much of what we claim to know is based on the word of authority. That fly agaric mushrooms are poisonous, the area of Sweden is 450,000 sq. km., the last ice age ended 10,000 years ago, *On Virtues and Vices* is a spurious work of Aristotle, the compositional formula for water is H_2O, King Erik XIV of Sweden died in 1577, plutonium is radioactive, some noble gases form compounds, nothing travels faster than light, the equation $x^n + y^n = z^n$ has no non-zero integral solutions for $n > 2$, and so forth, are facts for which most of us have no direct evidence but which we nevertheless accept on the strength of the authority of our teachers and textbooks.

This can easily lead us astray. A well-known American philosopher with an amateur interest in geography tells the tale of once walking around Monaco and remarking to his companion, "Just think—only eight square miles!" (Quine and Ullian 1978, p. 56). His sceptical companion thought even eight square miles was a considerable exaggeration, and his doubts were soon confirmed by a consideration of the map. Yet *Encyclopaedia Britannica*, our philosopher's source, gives the area as approximately eight square miles. Moreover, the World Almanac, Scott's stamp album, various American atlases and the gazetteers in American dictionaries all agreed on eight square miles. Hachette and Larousse, on the other hand, agreed rather on something less than three fifths of a square mile. Closer examination of *Encyclopaedia Britannica* (11th ed.) brought to light the alarming statement "Area about 8 sq. m., the length being 21/4 m. and width varying from 165 to 1100 yds.".[2] Had the editors of *Encyclopaedia Britannica* done their arithmetic correctly—and in this case, it suffices to note that multiplying 21/4 by a fraction reduces it, and cannot possible increase it to 8—the mistake would have been nipped in the bud. What is worse is that several other reference works presumed they could rely on the authority of *Encyclopaedia Britannica* without the need for an independent check, even when the grounds on which the original inference should have been made were clearly displayed. We might as well try to confirm the headline story in our daily newspaper by buying more copies of the same paper. Seeking safety in numbers by checking several sources is a sensible precaution, but comes to nothing if the different sources are not independent. Like the latest fashion, opinions may become trendy as more people concur without gaining anything in reliability.

---

[2]There are 1560 yards to a mile.

Though perfectly innocent, a false opinion may gain the status of authority through misjudgement or mistaken reasoning. More sinister motives lurk behind misleading stories put about by advertising agents trying to persuade us to buy remedies for nonexistent maladies, or politicians trying to gain support for a change of social or foreign policy. Whatever the interests at stake, the trust we put in the testimony of others is related to the possibility of confirmation should we wish to take the trouble. Our suspicions are raised when the very possibility of an independent check is ruled out. The perfectly innocent member of the public who is completely unaware of the source of the piece of information he is promoting—who can't remember first reading or hearing it—is not likely to gain our confidence. Nor is the dogmatist who insists we believe it merely because he tells us. The person who invites us to check for ourselves is more convincing. But even here we must be aware that this can be used as a device for gaining our confidence where the expectation is that we won't take up the invitation. Transparency is of the essence. Writers are expected, as a matter of course, to give their sources in intelligible form, following one or other recognised system of reference. Speakers are expected to volunteer their sources on request.

Students and laymen are by no means alone in their plight. Despite its drawbacks, reliance on the testimony of others is a necessity for gaining much of the kind of knowledge we seek. Historians and jury members have to rely on the accounts of witnesses. In favourable circumstances, their testimony may be independently corroborated by witnesses whose testimony is genuinely independent—i.e. it doesn't rest on hearsay going back to the original witness, but is based on their own judgement and powers of reasoning. The thought here is that it would be too much of a coincidence for independent witnesses to agree. But such an outcome is unlikely only provided that the matter at issue is sufficiently out of the ordinary to minimise the chances that witnesses can be reasonably expected to independently guess at the same outcome. The vaguer the descriptions, the more likely independent witnesses are to agree. Witnesses familiar with a regularly occurring kind of phenomena might be expected to notice matters of detail that distinguish individual occurrences, and the nitty-gritty of corroboration of independent testimonies is sought in the finer points of the descriptions offered.

In less favourable circumstances, or as a separate control on the veracity of independent witnesses, the question of general trustworthiness may be raised. A person might be known as an inveterate liar, and that settles the matter. But frequently, the empirical record of the witness's veracity is unavailable or of little bearing in the situation at issue, and we have to resort to what we know about psychological motives. What are the witness's own interests in the matter at hand? Testimony is more plausible, the less influence it has on the comfort and well-being of the witness, or of how it will reflect on the evaluation, moral or otherwise, of the witness. Or at any rate, where there are such relationships, a testimony that reflects negatively on the witness's own person is presumably one he would prefer to conceal, and is all the more plausible for that reason. The descriptions of the events of a certain epoch that have come down to us through history may be dominated by the desire of the historical witnesses to present themselves and their own deeds in a positive light and others negatively. In his own account of his reign, King Gustav III of Sweden paints a picture of conflict during the Period of Liberty which he describes himself as quelling. He presents himself as a unifying light, striving, after the coup d'état restoring the power of the throne in 1772, to bring the divisive estates of the realm into agreement, which the country so badly needed. This self-serving description is, of course, taken by historians with a nip of salt. The varying newspaper reports of the rowdy student manifestations in Uppsala after the French revolution in 1848 are naturally understood as the products of the liberal and the conservative press.

Establishing a text sometimes raises a problem not immediately resolved by the available "witnesses" in the form of presently existing documents. None of the original documents of Aristotle's texts has survived, and scholars have to rely on sources deriving from the originals by a long line of intervening circumstances, including copying of the original Greek manuscripts, assembling copies in the library at Alexandria several centuries later, translation into Latin and Arabic, retranslation from Arabic into Latin, and discoveries of Greek copies as the works were reintroduced into Western Europe. The modern scholar is faced with a variety of sources which are each the result of a line of copies reaching back to the original documents. Many links are missing, and many lines converge. Comparison of copies helps to identify a mistake or deliberate addition which has been "inherited" by subsequent copies. Two such additions in a copy, other things being equal, would argue for it being later than a copy with only one. Such clues can help structure the variations into a tree all of whose branches go back to the original, and some of which converge with others in the later than direction where a copy derives from several others. Several works attributed to Aristotle in a long tradition of study are certainly or probably not written by Aristotle. The spuriousness of some others is seriously doubted, often on internal grounds connected with the interpretation of the presumed corpus. This provides a testing ground for some points of interpretation and some of the issues may well never be finally settled.

Acquiring and sustaining a body of knowledge is often a corporate venture, in which the claims made by one person rely on the authority of others. Experts in one field are laymen in others, and will rely on the authority of other experts when their enquiries encroach on adjoining fields. Projects are undertaken by large groups of investigators, none of whom masters the entire range of relevant disciplines. Failure to take cognisance of different fields of knowledge may well delay or completely block the acquisition of knowledge. After the first alarms about the depletion of the ozone layer were sounded in the 1970s, three hypotheses were initially developed. Chemists thought it was a purely chemical matter of the production and consumption of chlorine radicals. Meteorologists, on the other hand, thought that lower atmosphere air containing little ozone welled upwards due to a major climatic shift in 1976. Finally, physicists saw solar activity as the cause of the depletion. These alternatives were clearly the products of different disciplines, whose practitioners were unfamiliar with the ideas and methods of the other disciplines and initially unwilling to countenance their relevance. The later development of the chemical thread was prompted by chemists' communication with meteorologists, who were able to point to the presence of stratospheric "mother of pearl" clouds made up of ice crystals formed in the cold winter and surviving into the spring. Previously, chemists thought of the atmosphere as a paradigm environment for purely homogeneous gas phase reactions. But some special factor was needed to account for the regeneration of chlorine radicals in the chain reaction destroying ozone. The expected source familiar to chemists derived from ultraviolet light. But this is clearly absent in the early Antarctic spring, yet the phenomenon of ozone depletion was restricted to the Antarctic. The solution was to be found in the possibility of a heterogeneous reaction on the surface of ice crystals in polar stratospheric clouds. These do form in the Arctic too, but in smaller numbers, and disperse before spring.

Statisticians are frequently called upon to advise on aspects of experimental procedure or analysis of data. Reliance on others is a notable feature even within single established disciplines, however. This is not just a matter of finding more hands to work with and more eyes to see with. The detailed consideration of sources of systematic error appeals to the whole body of previously established science. Laws discovered by our forefathers warn of the side-effects of phenomena arising in the experiments we conduct today. But more broadly, scientists pushing back the frontiers of knowledge stand on the shoulders of others, in Newton's phrase (quoted in Sect. 3.4), not only in critically evaluating results and searching for sources of error. The general ideas underlying the projects they propose to undertake are evaluated in the light of the present state of knowledge—i.e. in the light of what others have done in creating the present state of knowledge. The current state of knowledge determines expectations and what is of interest in what remains to be explored. Of course, new results may conflict with the prevalent opinion, calling for further investigation to decide the matter. Claims to knowledge are fallible. The possibility of future conflict can never be completely ruled out, and what seems at one time to be a well-justified claim may have to be qualified or withdrawn later. But the investigator proceeds on the basis of his present understanding, and if the results of his efforts are to be valued by the intellectual community then the investigator's present understanding should be based upon a thorough and detailed acquaintance with all the relevant literature. The difficulties of commanding such an overview imposes the discipline of specialisation, with the attendant risks noted above. But rediscovering the wheel, however exciting the experience, makes no impression on the cognoscenti. Similarly, criticism of extreme or outdated versions of the antagonist's position might be standard fare in politics, but such cheap manoeuvres are not acceptable in scientific debate. In his recent book, *Häxornas Försvarare* (2002), Guillou scores some easy points against the once popular theory from the 1970s that witches were persecuted because they posed a threat to the emerging male profession of physicians. But this is too narrow and outdated a critique on which to base a reasonable assessment of feminist theories. The objective status of knowledge is not merely a matter of satisfying a philosophical ideal of presumptive universality—validity for all men. There is the concrete, practical requirement of satisfying a peer review.

2.7 Error, Risk and Values

Contrasts between what is internal and what is external to science, and between what is a matter of fact and a question of values, may seem clear when raising more grandiose issues of the import of science on society in very general terms. But these distinctions are less readily made at the coal face, when it comes to the nitty-gritty of assessing the import of scientific work on matters of immediate human interest such as health and the environment.

It is, perhaps, a common view that science is concerned with establishing facts which are entirely independent of questions of value and moral appraisal. As we will discuss in Chap. 4, scientists might be faulted for not giving as accurate and precise an account as possible of the known facts. They have a moral duty to take reasonable steps to facilitate a free and open discussion of the issues, to act responsibly in the light of the likely bearing of scientific knowledge on human welfare and to cooperate with social, political and commercial interests appropriately. In addition to determining whether the development of new technologies is desirable, values must be consulted in determining which problems to pursue and how they should be pursued (by respecting human integrity, animal rights and so on, even if this limits the attainment of knowledge). But the facts themselves, on this view, are independent of all such considerations of value. To allow values to encroach upon the description of the facts would threaten the objectivity of science and impair decision making, which should be based on clearly distinguished statements of the relevant facts (or rather, their probabilities) and the relevant values (or rather, a systematic assessment of the relative value of the various desirable goals).

Against this view, Richard Rudner (1953) argued that values enter the internal reasoning of science. It has been emphasised earlier that a reasonable attitude towards science and scientific procedure is governed by its inductive character. It is a damaging and misleading mistake to describe its achievements as scientifically proven. Results are not established in science as Euclid thought his theorems were demonstrated, by unassailable reasoning from indubitable premises; they are more or less strongly supported by the available evidence. History bears witness to countless theories and hypotheses once sternly advocated by first-rate scientists but since shown to be wrong. The appropriate scientific attitude is to modestly believe at any given time in theories firmly supported by the currently available evidence which nevertheless falls short of proving them, always aware that the future may bring counter-evidence. Rudner's point is that

> since no scientific hypothesis is ever completely verified, in accepting a hypothesis the scientist must make the decision that the evidence is sufficiently strong or that the probability is sufficiently high to warrant the acceptance of the hypothesis. (Rudner 1953, p. 2)

Thus, when accepting a theory or hypothesis as true on the basis of the available evidence, there is a certain possibility that a mistake has been made and it is not true. Similarly, there is a certain possibility that hypotheses or theories dismissed on the basis of the evidence are in fact true. By setting standards high, the risk of accepting as true what is in fact false, and believing false what is in fact true, is reduced. But the possibility of error cannot be eliminated altogether, and how far the standards can be improved is limited by the available resources. The consequences of possible error guide the setting of the standard.

In pure science, the consequences are evaluated in terms of what might be called epistemic values such as the reliability, extensiveness and systematic character of what we claim to know (Hempel 1965, p. 93), which must be balanced against the resources available. Sharply contrasting epistemic values were at work on opposite sides of the phlogiston debate, which serves as a striking illustration. Antoine Lavoisier (1743–1794) set great store by simplicity and elegance of theory, whereas Joseph Priestley (1733–1804) thought completeness was more important. Accordingly, Priestley and fellow phlogistonists sought to explain all the phenomena they observed and described, even at the cost of making their hypotheses messy and inelegant. Lavoisier and his followers repeatedly cited the simple and straightforward paradigm case of heating mercury in air to produce the red calx of mercury, which when heated more strongly would emit a gas (variously called oxygen or dephlogisticated air), neatly illustrating the processes of oxidation and reduction (de-oxidation). But Priestley took the view that focusing on one exceptional case distorted the whole picture, because other metals behaved differently. The oxidation-reduction cycle could not be performed on iron, for example, whose calx could only be reduced when in contact with some other substance, which Priestley supposed must contain phlogiston. Lavoisier protested against the growing complexity of the phlogistonists' theory and conflicting changes they introduced, searching himself for commitment to a more permanent and elegant theory. Priestley, on the other hand, reconciled himself to frequently changing his theory, and wrote that the pursuit of truth should make the experimentalist wary of acquiring too great an attachment to hypotheses. He didn't deny the value of simple theories, just as Lavoisier didn't deny the value of completeness. But they had different views on the relative importance of these values, which played a role in their scientific judgement when they came into conflict (Chang 2012b).

Non-epistemic values may come into play where the hypothesis or theory in question impinges on the interests of a certain group, and the consequences of possible error must be taken into account in the evaluation of the possibility of erroneously accepting or rejecting the theory. Rudner continues the passage quoted above as follows:

> Obviously our decision regarding the evidence and respecting how strong is "strong enough", is going to be a function of the importance, in the typically ethical sense, of making a mistake in accepting or rejecting the hypothesis. Thus, to take a crude but easily managable example, if the hypothesis under consideration were to the effect that a toxic ingredient of a drug was not present in lethal quantity, we would require a relatively high degree of confirmation or confidence before accepting the hypothesis—for the consequences of making a mistake here are exceedingly grave by our moral standards. On the other hand, if say, our hypothesis stated that, on the basis of a sample, a certain lot of machine stamped belt buckles was not defective, the degree of confidence we should require would be relatively not so high. *How sure we need to be before we accept a hypothesis will depend on how serious the mistake would be.* (Rudner 1953, p. 2)

Philosophers have responded by arguing that such considerations are a matter for those applying the theory in question and not for those investigating its truth (e.g. McMullin 1983). But more recently, writers such as Carl Cranor (1993) and Heather Douglas (2000, 2004) have taken up the banner and argued that potentially damaging effects, should the scientist be wrong, are sometimes not brought to light in public decision making because inductive risks have been taken prior to acceptance or rejection of theories ostensibly based on purely epistemic values. The following discussion draws heavily on Douglas (2000).

Dioxins are a class of chemicals discovered in the late 1950s which are produced as a by-product of industrial processes such as pesticide production, paper bleaching and waste incineration. They are also produced naturally from volcanoes and forest fires, but their level in the biosphere has increased dramatically in the twentieth century. Their acute affects on human health are well-known from industrial accidents such as that in Seveso, Italy, in 1976. Studies are usually conducted on the most toxic dioxin (Kociba et al. 1978), here simply called dioxin, and experiments show that it kills rodents and primates even at relatively low doses. But more long-term effects such as cancer are more uncertain. Despite extensive studies since 1970, the scientific Advisory Board of the Environmental Protection Agency in the U.S.A. failed to reach agreement in Spring 2001 on the extent of the risk posed by dioxins. This kind of disagreement is not uncommon on environmental issues where there are very many factors at work and the complex causal mechanisms are difficult to fathom. Something is known about how dioxin enters the cell and makes its way to the cell nucleus. But there it interacts with many genetic sites, and the mechanisms become overwhelmingly complex, presenting scientists with an explanatory black box. Calling for a full understanding of the biochemical mechanisms before imposing regulation on industry is clearly a disingenuous delaying tactic given what is already known about its effects on whole organisms. The question is how to set levels in regulating industry in order to provide a reasonable degree of protection of public health.

The effect of different levels of dioxin on the inducement of cancer is investigated by considering how different groups of rats subjected to different doses fare in comparison with a control group over a period of time approaching the typical life-span of a rat. Because of the extreme costs and difficulties of running studies on larger numbers, current studies typically use between 50–100 animals in each dose group. Ethical and economic values clearly have to be balanced in determining the detailed character of studies undertaken, but that is not the present point. Once the design of the study or experiment is settled, further manipulation of the potential error is only possible by trading off the risk of mistakenly accepting false hypotheses against the risk of mistakenly rejecting true hypotheses. In setting the standard of statistical significance, a stricter standard reduces the rate of acceptance of false hypotheses at the cost of increasing the rate of rejecting true hypotheses, and vice versa. Thus, with a laxer standard of statistical significance, a smaller difference in cancer rates between the exposed and the control populations will be considered significant, reducing the risk of claiming that a given level of dioxin doesn't cause cancer when it in fact does. At the same time, the risk of giving the false appearance that dioxins cause more cancer than they in fact do is increased. If heeded by the authorities, this will lead to over-regulation of the chemical industry, resulting in excess costs and the social disruption caused in its wake. But this might be thought preferable to an excess of rejected hypotheses which are in fact true, giving the impression that dioxins are less harmful than they actually are and leading to an under-regulation of the chemical industry. The costs of this strategy would be borne by a worsening of public health and its affects.

The evaluation of these alternatives clearly appeals to values over and above purely epistemic values, and will determine which balance between accepted falsehoods and rejected truths is preferable. Of course, the standard of statistical significance might be set without consideration of such values. But why should disregarding relevant considerations be thought to give a more objective account of the matter? A choice of standard will correspond to a certain position on the preference of these alternatives, whether we are aware of it or not. Presenting the results of an investigation into the health risks of dioxins as though it presupposes no such evaluations is merely misleading by reason of omission. That would be morally reprehensible on the part of intelligent scientific advisers, and should not be hidden behind a facade of objectivity. Political decisions should be made on the basis of complete transparency. Any balancing of the concerns about public health and dioxin pollution against those of protecting industry from increased regulation should be brought into the open.

Setting a standard of statistical significance is a methodological issue. Non-epistemic values may also enter into the interpretation of the data provided by empirical investigations in ways other than those entering into considerations of methodology. In the rat studies, data is provided by postmortem examinations of the animals at the end of the experiment. A particular study, conducted by Richard Kociba and other toxicologists (Kociba et al. 1978), has played a key role in determining acceptable levels for dioxins in the environment. This study focused on cancers of the liver in female rats, and the liver slides produced in this study have since undergone two further examinations by pathologists in 1980 and 1990. Although the same slides are common to all three examinations, no two of the evaluations agree. This disagreement among the experts suggests that determining whether a sample exhibits cancer is not straightforward. The uncertainty carries inductive risk and the consequences of possible errors must be evaluated. How, for example, should the borderline cases be treated? Classifying them as non-cancerous reduces the acceptance of false hypotheses while increasing the likelihood of rejected hypotheses being true. This strategy will underestimate the risk to health. Accordingly, relaxing the regulation of dioxin leakage as a result of a reinterpretation of the data (as happened in the state of Maine in the U.S.A. after the 1990 reevaluation) may lead to increased harm to the public. Thus, the pathologists who judge the borderline cases to be non-cancerous must also judge this to be an acceptable risk. They are not taking a neutral position on this issue of values. Conversely, classifying borderline cases as malignancies reduces the risk of rejecting truths at the price of increasing the likelihood of accepting falsehoods. This strategy increases the apparent incidence of cancer in rats, and the apparent risk to health, which should lead the authorities to impose more stringent regulation of the chemical industry. Better protecting public health in this way is thus achieved at the economic cost of possibly unnecessary regulation.

Douglas (2000) has an interesting discussion of the use of blinding the pathologists so that they are ignorant of which dose group the slides belong. The thought is that this would counter any tendency on the part of a pathologist to avoid accepting false hypotheses by rejecting more true ones, or vice versa, by evenly distributing that tendency among control and dosed slides. If the number of borderline cases were the same in each group, the difference between control and dosed slides would be roughly the same. This blinding methodology would prevent an unacceptable direct use of non-epistemic values in the classification of data. Unfortunately, in the present case, complete blinding is not possible because, of the three dose levels used in the experiment, the two higher levels lead to visible signs of poisoning in the liver cells with which the pathologists are familiar. Most of the borderline cases occur at these higher levels where the slides can be identified, implying that the pathologists were aware of the values underlying their classification of the slides.

It is not realistic to think that issues like this one of fixing a reasonable degree of regulation on industry can be settled in the near future in simple empirical fashion, investigating whether cancer rates fall in a given period when regulations are tightened. There are so many additional factors which might account for a drop in cancer rates—from changes of diet to changes in exposure to environmental features—that a change in regulation cannot be isolated as the cause. Again, any reduction in the incidence of disease might not be detectable by available methods, for statistical or other reasons. The world is too complex to provide an experimental basis on which to test hypotheses guiding environmental policy. Politicians must resign themselves to this situation and not press for simplistic guidelines. At the same time, scientists cannot remain silent in the face of all this uncertainty. Since they are the ones in the best position to make relevant value-loaded judgements, then they shouldn't shrink from doing so. Given the vast increase in public expenditure on scientific research during the last half-century, it is perhaps not unreasonable that the general public should expect this of them where the likely import of error impinges on public welfare.

2.8 Summary

Raw data, unsoiled by human contact, is seldom of scientific interest. To the extent that independence of the observer is a criterion of objectivity we should strive to meet, it is not a matter of being completely observer-free. Without the intellectual faculties of the observer—perception and understanding—there would be no data. The notion of independence is rather that of being independent of any specific observer. Anyone can—or where repeatability is, of the nature of the case, impossible, could—perform the observation. This requires that bias and preconceived ideas be eliminated, or at least accommodated. And this in turn requires that the various kinds of personal contribution or interference be recognised. The critical study of conditions under which investigations are conducted may well take us beyond the mental presuppositions of the investigator, and require a detailed examination of the circumstances in which experiments are conducted, observations performed and data collected. General rules guiding such critical reflection applicable to all disciplines are hardly to be expected. But students are given a grounding in the relevant kind of considerations in the course of studying particular disciplines.

Guarding against error is a matter of recognising and restricting the risk of error, with a view to providing a quantitative or a qualitative formulation of the remaining margin of error. Exactly how this degree of approximation of results is conveyed in the formulations of knowledge claims is also very much a matter for individual disciplines. There is a great deal that is tacitly taken for granted in the literature, and so much that might be missed by the casual reader for whom scientific publications are not primarily intended. But it is naive to think that explicit reference to degree of approximation is a sign of vagueness. On the contrary, it is a mark of precision of thought, whose articulation requires all the skill acquired in years devoted to studying an established discipline. An apparently concise formulation making no reference to limits of probable error reveals unawareness of problems which would be faced by a more critical study and may, as we saw in the last section, even involve value judgements. Inability to enter into discussions of such matters is a mark of the amateur. It is unfortunate that so much of the popular reporting of science totally ignores this aspect in pursuit of deceptively simple sound bites and unedifying analogies. Concise grammar may be necessary for concise thought, but it is certainly not sufficient for precision and accuracy. Unrefined reporting of what has been scientifically proven is more reminiscent of the worst kind of uncritical religious dogma than the reasoned deliverances of an objective assessment of the evidence. We have seen that fallibility is the inevitable consequence of the process of acquiring precise knowledge. The scientist doesn't guard against error by becoming certain, and can confidently make a well-justified claim to know because there is no requirement of knowing that one knows.

The role of others has been emphasised in establishing intersubjectively guaranteed standards of objectivity. We rely on the testimony of others for much of our knowledge in the form of the observations of witnesses and the authority of teachers, reference books and textbooks. But sources should in principle be susceptible to critical scrutiny, and at the very least be in principle checkable by clear statements of grounds and sources they in turn rely upon. Building on and critically assessing the work of others can, over a period of time, lead to progress in the form of the development and refinement of concepts crucial to the articulation of general laws.

Chapter 3
Countering Relativism

3.1 Vanquishing Reason

It was argued in Chap. 2 that aspiring to endow our beliefs with the status of knowledge calls for a justification meeting appropriate standards of objectivity. Which of the multifarious criteria are applicable in any particular case depends very much on the matter in hand. But the attendant circumstances contrive to remove any possibility of absolute certainty. The general point to emerge was that the naive view, that any intervention of the observer detracts from objectivity, is completely untenable. On the contrary, the investigator must be able to accommodate circumstances in reasonable fashion with a view to bolstering the claim so that it will hold up to public scrutiny. For although knowledge is inevitably possessed by an individual, it can also be shared by many individuals, and any knowledge claims we make are presumed to be knowledge claims that anyone else might make their own. If there is a perspective imposed by one individual, then it is a perspective that can be shared by any individual. Although difficulties with justification create complications, they do not undermine the very idea that knowledge is objective, and something we may reasonably strive after.

Another line of objection challenges the possibility of acquiring a broad and substantial body of knowledge by questioning the notion of truth rather than justification. Truth, as we have seen, is a necessary condition of knowledge. If a researcher knows that such-and-such is the case, then such-and-such is the case. Should further research reveal that such-and-such is not the case, then it follows, not that there is no such thing as knowledge, but merely that our researcher did not know that such-and-such after all. He thought he could reasonably claim to know such-and-such, although it transpired that he was wrong. Were it to be shown, not merely that some particular claim is not true, but that there are no truths, however, then it would ipso facto be impossible to have any knowledge. Startling as it is, this claim is made in one way or another by authors who maintain that truth is relative to the beholder. It calls into question a presupposition of the earlier discussion, that

© Springer Nature Switzerland AG 2020

P. Needham, *Getting to Know the World Scientifically*, Synthese Library 423,
https://doi.org/10.1007/978-3-030-40216-7_3

objective judgements are valid for all men. Judgements can in principle, it was said, be shared by all men and all men can, in principle, agree upon them. Not so if there is anything to the present objection.

A view which questions the idea that truth is truth simpliciter, without qualification, is called relativism. This is a general term for many evils. One simple-minded expression of the relativist doctrine has it that truth is relative to a person. There are no truths simpliciter, only what is true for you, or true for me. This idea sometimes arises where there is controversy, in the vain hope of resolving, or at least eliminating, unpleasant disagreement by redescribing what John thinks and what Sally thinks as what is true for John and what is true for Sally. This device allows us not to agree to disagree, but more accurately, not to disagree, so we needn't come to blows. There is a price to be paid for eliminating disagreement in this way, however. It makes agreement impossible. It may well be so that water is wet for John and water is wet for Sally. But since truth is, by hypothesis, always relative to a beholder there is no inferring from these two facts that water is wet. (Just two facts? Isn't the fact that water is wet for John itself an absolute truth, and so should give way to "water is wet for John for Tom" and so on for Dick and Harry? But why stop there? Surely "water is wet for John for Tom" is no better than "water is wet for John", and we should affirm water is wet for John for Tom for Harry, etc. Relative truth is truly ineffable—that is, ineffable for each of us ....) Normally, to say that two people agree is to say that they agree on something—that something is the case. Clearly, resolving disagreement by the true-for-me-or-you ploy is not finding agreement but merely dropping hostilities.

Perhaps the thought behind the original suggestion is really no more than an expression of the fact that different people often believe different things. To say that there is a ghost in the attic is true for John but not for Sally is just to describe two people's divergent beliefs. But then we already have a perfectly adequate word for the attitude of acceptance an individual adopts towards a proposition, and in serious talk there is no need to misuse the word "truth" to express what is adequately conveyed by the word "belief".

Is the thought behind relativism that interests usually direct the search for truth? Interest is clearly relative and one person's interests may differ from another's. Does this mean that the truth is tainted with particular interests? Kepler's interest in Copernicus' theory was driven by his involvement in neoplatonic and hermetic movements of his day. Nineteenth-century social thought in Britain facilitated the initial acceptability of Darwin's concept of the struggle for survival. Without these positive predispositions towards Copernicus' and Darwin's theories when they were new and unproven, who knows what might have happened to them? Perhaps their opponents could have successfully ignored them and they might have sunk into oblivion. But just as people sometimes speak of truth for dramatic effect when what they are really talking about is belief, so we should be careful not to confuse what might be said about interests with what can be said about truth. Differences in socio-economic background may result in differences in two historians' interests, for example, leading them to investigate different circumstances connected with the growing population in Victorian England. They might disagree on principal causes, but the facts on which they each build their case are not relative to their respective interests. And where they do dispute the facts, the issue is the truth, not truth-given-such-and-such-interests. In general, the discussion of experimental results, interpretations of documents, competing theories, and so forth—in short, a critical debate—is a healthy state of affairs, leading to progress and interdisciplinary connections, all of which presupposes a shared concern with the same world.

Relativism is not a new doctrine. It has figured in our recorded intellectual history since the time of the ancients, and was recorded in Plato's *Theaetetus*, where Socrates is portrayed as challenging Protagoras' claim that "man is the measure of all things". Traditionally, the relativist is held to face a dilemma: The relativist's thesis, that all claims are true in some perspectives and false in others, is either (i) absolutely (without reference to any perspective) true or (ii) it is only true relative to the relativist's perspective. But the first alternative amounts to abandoning the thesis and the second gives the relativist's opponent no reason to accept the thesis.

The term "absolute" with which relativism is contrasted can lead to confusion. Relativism is the doctrine that nothing is true *simpliciter*, but always true relative to a person, a group of people, a society, a perspective, or whatever. This is opposed to the view that something is true without qualification, or if not, then it is simply false (in accordance with the law of excluded middle of classical logic). To add a qualification and speak of absolute truth is in fact redundant: absolute truth is simply truth. Redundancy is harmless enough, perhaps, and speaking of absolute truth might provide emphasis, serving to recall the opposing stance. But there is a danger of confusing the contrast with relativism with something else. Thus, the Swedish historian Knut Kjeldstadli maintains against the relativists that "certain theories are false and can be falsified", but nevertheless rejects "the belief that there is always just one truth, that which can be called an absolute standpoint" (1998, p. 282). What he presumably intends to say is that there are several relevant factors that bear on the issue in question, and it would be a misleading oversight to take only one into consideration. There is nothing intrinsically wrong in introducing the term "absolute" in ways not strictly confined to a denial of the relativist's thesis. But the danger arises that what several authors deny in rejecting absolutism need not be the same thing, and it is not immediately clear whether they stand in conflict with those who deny relativism.

The effects of differing circumstances may prompt the device of relativisation. Consider the fact that people with a tendency to accumulate lactose in the intestine, with resulting diarrhoea and abdominal cramps, are described in Northern Europe as suffering from a genetic disease. The symptoms are induced by consumption of milk in people who cannot produce lactase in sufficient amount. In Africa, on the other hand, where lactase deficiency is fairly common and milk consumption rare, people suffering from the same symptoms after drinking milk are said to be suffering from an environmental disease. Diseases are classified by their causes. Unlike the situation in Africa, lactase deficiency in Northern Europe is rare and milk consumption common, so what is unusual there is counted as the cause of the symptoms and the disease is classified as genetic. By the same count, the unusual feature in Africa of drinking milk is there counted the cause of the disease, which is accordingly classified as environmental. Is this situation best described by saying that it is true for Northern Europeans that these symptoms are due to a genetic disease whereas it is true for Africans that these symptoms are due to an environmental disease? Surely it is less obscurantist to say that, since exactly the same circumstances give rise to what is variously described as a genetic and an environmental disease, it is the distinction between genetic and environmental diseases, rather than the notion of truth, that needs some qualification.

Sometimes, different people mean different things by the same words. John might use "water" in the sense of the *Shorter Oxford Dictionary* (3rd. ed., 1992), namely "The liquid of which seas, lakes, and rivers are composed, and which falls as rain and issues from springs. When pure, it is colourless (except as seen in large quantity, when in has a blue tint), tasteless, and inodorous". Sally, on the other hand, influenced by her school chemistry, knows that water is a compound composed of hydrogen and oxygen in the proportions represented by the compositional formula H_2O and consequently realises that both ice and steam are water, which, because they are not liquid, John denies. The situation might be colloquially described by saying that ice and steam are water for Sally but not for John. The superficial relativisation is easily resolved, however, as a difference of usage rather than a difference of perspective, which can be straightforwardly described without resort to gestalt shifts by defining John's sense of water as water in Sally's sense confined to the liquid phase.

Change in the meaning of terms with new discoveries is not uncommon in the history of science. An acid was originally just a corrosive substance. Robert Boyle (1627–1691) defined it as a substance with a sour taste that could turn litmus red. A century later Lavoisier discovered that when various elements such as carbon, sulphur and phosphorous burn in oxygen they produce oxides which form acids when dissolved in water, and a defining criterion became that acids contain oxygen. It came to light that this excluded hydrochloric acid, which is a corrosive substance in the everyday sense and satisfies Boyle's criterion. Lavoisier drew the conclusion that hydrochloric acid must contain oxygen, but this was disproved by Davy and Lavoisier's definition in turn gave way to another according to which an acid contains hydrogen which can be released by replacement with metals. Towards the end of the nineteenth century, Arrhenius' theory of the dissociation of electrolytes led to a definition in terms of hydrogen ion concentration. This in turn was generalised in the 1920s by the notion of a Brønsted-Lowry acid, which is defined as a proton (i.e. hydrogen ion) donor (and a Brønsted-Lowry base is a proton acceptor). At the same time Lewis generalised the notion of an acid in different way, and a Lewis acid is defined as an acceptor of electron pairs.[1] There is nothing sacred about modern science. Even relatively new terms in the vocabulary of science are not insulated from well-motivated modification.

Does the relativism dismissed above reappear in historical guise? What was an acid for Boyle is not an acid for us, and vice versa. This is not true. Whatever Boyle counted as an acid so do we. On the other hand, there are many substances

---

[1] See Chang (2012a, pp. 693–5) for an account of the difference between these two concepts.

discovered since Boyle's time that we count as acids on which Boyle obviously didn't have an opinion. Surely there's no case to be made for relativism on the basis of Boyle's understandable ignorance.

But not everyone agrees that the threat of relativism posed by the phenomenon of change of meaning can be averted by discerning ambiguity and understanding that concepts develop in a successive, step-wise manner. Such devices have not satisfied the historian Thomas Kuhn (1922–1996) and the philosopher Paul Feyerabend (1924–1994), who have each argued for a more radical view in the light of change of meaning in the course of the history of science. Feyerabend's argument is taken up in a later chapter (Sect. 6.6). Kuhn is discussed in the next section.

3.2 Radical Meaning Change

In one passage, Kuhn states his position as follows:

> … the proponents of different paradigms practice their trades in different worlds. … Practicing in different worlds, the two groups of scientists see different things when they look from the same point in the same direction. … That is why a law that cannot even be demonstrated to one group of scientists may occasionally seem intuitively obvious to another. Equally, it is why, before they can hope to communicate fully, one group or the other must experience the conversion that we have been calling a paradigm shift. Just because it is a transition between incommensurables, the transition between competing paradigms cannot be made a step at a time, forced by logic and neutral experience. Like the gestalt switch, it must occur all at once (though not necessarily in an instant) or not at all. (Kuhn 1962/1970, p. 150)

Writing with the authority of a historian of science, it seems that Kuhn has undercut in one fell swoop the possibility of rational debate on the truth. Appealing to "logic and neutral experience" is of no avail. Investigators pursuing different theories inhabit "different worlds". What is obvious to the one group is a mystery to the other. Their views are "incommensurable", and there is no possibility of understanding what the other says short of "conversion" to the other side. The eighteenth century Scottish philosopher David Hume (1711–776) understood that having a belief is more than having an idea when he realised that we distinguish between entertaining another's thought and actually believing it.[2] Engaging in meaningful dispute with someone of a different opinion would be completely pointless if there were no possibility of grasping another's view without going so far as to accept it. But precisely this distinction is obliterated by Kuhn's notion of a "paradigm shift".

Kuhn raises two problems. Can we claim to understand thoughts and actions of agents in the past when these agents had beliefs different, perhaps radically

---

[2]"Suppose a person present with me, who advances propositions, to which I do not assent, *that* Cæsar *dy'd in his bed, that silver is more fusible than lead, or mercury heavier than gold*; 'tis evident, that notwithstanding my incredulity, I clearly understand his meaning" (*Treatise* I.III.vii; Hume 1739, p. 95).

different, from ours? Is it possible for contemporary adherents of different theories or views, or members of different cultures, to understand one another while retaining sufficient distance to fairly assess the truth of the claims they make? The presuppositions necessarily involved in the one party's taking a stand on an issue or adopting a theory led Kuhn to give a negative answer in both cases.

Whatever the detailed reasons that might be offered for such a stance, it is odd that Kuhn should adopt it in his capacity as a historian. The argument he develops in his book, *The Structure of Scientific Revolutions*, rests on a number of cases taken from the history of science which he presents with the authority of a historian. How does he expect his reader to take this argument? The historical studies are the evidence on which he rests his case. If this evidence is questioned as a fair account of the historical truth, his case is undermined. But the conclusion he wants his reader to accept is precisely that any such notion of independently accessible truth is a chimera. On pain of inconsistency, he can only be hoping to surreptitiously convert his readers to the "paradigm" way, perhaps by fooling them into thinking that they are impressed by the weight of evidence and the force of the argument, when in fact he is subverting their rational and critical faculties.[3]

Those sceptical of this form of persuasion might seek a motivation for the relativist's thesis in the fact that the claims of the "foreigner"—the person from the past, or from the opposing camp, or the strange culture—are formulated in terms of theory-laden concepts, i.e. concepts which were introduced in connection with specific hypotheses or theories. What should we say about such concepts where the associated hypothesis or theory is called into question? This case arises often enough with the progress of science and improvement in general education, calling into question and rejecting theories once held in high esteem. It is not unusual that the associated concepts also fall by the wayside. We no longer speak of phlogiston, or caloric, nor of ether in the sense of an imponderable fluid pervading all space and supporting the motion of light. And although we still use the terms "earth", "water", "air" and "fire", they are no longer understood as elements in the way the ancient Greeks understood them. Nor do we think of health in terms of the balance of opposing fluids in the body which is upset in illness and has to be restored by draining blood. But if the meaning of terms is tied to theories to which we no longer subscribe, doesn't this mean that all claims made in these archaic terms can no longer be taken to be true? For to be true, a sentence must at least be meaningful, and precisely this prerequisite is undermined when the meaning-importing theory

---

[3] It has been suggested to me that for Kuhn, the scientific truth in question is supposed to be theoretical (i.e., claims about theoretical posits, such as quarks), whereas historical truth (i.e., claims about historical events, such as the founding of the Royal Society in 1660) is not. But Kuhn's historical evidence, as we will see, is the scientific case studies he elaborates, which are largely concerned with claims about theoretical posits and hardly at all with historical events like the founding of this or that institution. In any case, Kuhn has been an important source for the critique, widely accepted since the mid-1960s, of the observation-theory distinction introduced by the logical positivists. We will see Kuhn's view of the observation-theory distinction emerge towards the end of this section.

is rejected. Once we allow that the claims of our forefathers were "true from the perspective of the theory they upheld", although not true in the light of modern theory, however, then we seem obliged to say the same of ourselves. Surely the same fate will in all likelihood befall our own theories, together with the terminology used to formulate them, and they will in turn be superseded. So any claims we make about what is true and false are implicitly relativised to our present theoretical stance, and so not true or false *simpliciter*, but only from the perspective of modern theory. The fact that we don't usually express ourselves in this way is to be understood as a façon de parler—a convenient manner of speaking which conceals the (real?) import of what we say, but which is revealed by reflection of the present sort.

But we cannot let the premises of this line of argument pass so easily by without challenge. Before going into the matter in a little more detail, let us note that meaningfulness is every bit as necessary for falsity as it is for truth. In order to declare a sentence false, we have to know what it means. And even if the meaning of a term is determined by the hypotheses and general theory in the context of which it is used, this doesn't imply that the theoretical context *uniquely* determines the term's meaning. The term may well have the same, or much the same, meaning in a succeeding theoretical context. Disagreement over truth claims is not an insuperable obstacle to understanding other cultures. We no longer think it physically possible for people to walk on water and fly on broomsticks, so whatever the poor souls who claimed to witness, or confessed to actually practising, witchcraft, were really doing, we can't accept that they were telling the truth. Rather, we try to understand the phenomenon by offering alternative explanations—that the statements were elicited under hallucination, coercion or promise of reward, by people in authority irresponsibly encouraging impressionable youths in pursuit of their own interests, and so on. Rejecting the possibility of witchcraft is an important feature of our interpretation, but this disagreement wouldn't be possible if the confrontation of our beliefs with those of our forefathers could only be bridged by conversion, and if the comparison necessary for conflict couldn't be meaningfully articulated.

Let us consider one of Kuhn's principal cases, the phlogiston theory advocated by Georg Stahl (1669–1734) and subsequently developed by Priestley and others. The basic idea of this theory is that the combustion of a substance results in the release of phlogiston to the air. When wood burns, for example, phlogiston is released leaving ash as the (much lighter) residue. Similarly, when a metal such as mercury is heated, it exudes phlogiston leaving the calx of the metal. Nowadays we believe a diametrically opposed theory, due to Lavoisier, according to which combustion proceeds by absorption of a substance in the air we call oxygen, resulting in the formation of oxides. This doesn't prevent us from recognising a certain amount of truth in some of the accounts given by phlogistonians, however. For example, they knew, we would say, that combustion in a closed space ceases after a while. They explained this by supposing that air has a limited capacity for absorbing phlogiston. By heating the red calx of mercury, Priestley obtained the metal mercury and a new kind of air which he called "dephlogisticated air". For, so he reasoned, the calx of mercury had been turned into mercury by the absorption of phlogiston, which must have been taken from the air. Naturally, dephlogisticated air supports combustion and supports respiration (of mice, and himself, as Priestley observed) better than does ordinary air since removal of phlogiston from air leaves the air with a greater capacity for absorbing phlogiston.

Surely this won't strike the modern reader as an impenetrable morass of incomprehensible ideas requiring conversion to be understood. We can see a consistency in the line of thought developed to account for simple facts that we too seek to explain. Moreover, the idea that dephlogisticated air has a greater capacity for absorbing phlogiston is not completely off the mark, although we "wouldn't put it in quite that way". Surely the constructions with terminology now foreign to us don't hinder our recognising as true in this account the claim that the heating of calx affects air in such a way that it enhances combustion. The fact that the key terms of phlogiston theory are theory-laden with the false presupposition that something is released in combustion is not the insuperable obstacle that the relativists claim to a reformulation of the major tenets of that theory in language laden with a contrary theory.

If the relativist were right, however, then the phlogiston theory would not only be inaccessible to us. It wouldn't even have been comprehensible to Lavoisier, who can't be portrayed as arguing against it but merely speaking in a parallel world of his own oxygen theory instead. Whatever Lavoisier did, it couldn't be seen as mustering the evidence in support of what he took to be a better account, since any such comparison would, by the relativist's lights, be impossible. But this belittles Lavoisier's genius, which wasn't merely confined to the creation of a theory explaining combustion and a series of associated chemical reactions. Just as we can see how the phlogistonians thought, so Lavoisier mastered the views of his opponents. He understood the elaborations and small variations distinguishing them, which made the phlogiston theory considerably more complex than indicated here, and skilfully argued his case by devising experiments whose outcomes posed problems for the different versions of the theory. Equally, his antagonists understood him, and there is no question of their rejecting his new quantitative approach and refusing to engage in debate with him. All parties agreed to the principle of no loss of weight in conversion of reactants to products, and proceeded to argue in commonly accessible publications how their observations should be interpreted consistently with this principle. What clearer indication of a genuine debate could there be than the fact that *An Essay on Phlogiston* by Richard Kirwan, one of the leading phlogistonians, which appeared in 1783 was translated into French and published together with comments by Lavoisier and other French chemists, and these were in turn translated in a later English edition in 1789 together with Kirwan's replies? Lavoisier could identity what Priestley called dephlogisticated air with what he called oxygen, thus pinpointing the areas of disagreement. That is not say that this clinched the matter. There was no sudden and clear-cut end to the debate. Lavoisier's theory continued to attract criticism after his death and into the nineteenth century, not least because of the role played by caloric (Sect. 7.3). There is no good reason to mystify the new challenge as a mysterious gestalt shift of inscrutable paradigms.

In response to criticism for the unclarity of his critical apparatus (e.g. Masterman 1970), Kuhn responded in ways which didn't really satisfy his critics but gained him latter-day supporters. One such is Brad Wray, who gives the following account of Kuhn's revised scheme.

Essentially, what is involved is the introduction of changes in the scientific lexicon such that former relations of genus and species are no longer preserved. Instead, classes of objects that were previously regarded as related to each other as species to genus are no longer regarded in this way. For example, with the Copernican Revolution in astronomy the Sun was no longer regarded as a planet. Instead, it came to be regarded as a star. And the distinction between fixed stars and wandering stars was abandoned. ... The definition of planet [was also] changed from wandering star to satellite of the Sun.

Kuhn contrasts theses sorts of changes to a scientific lexicon with changes that merely enlarge a scientific lexicon. ... For example, when a new species of frog is discovered the class of amphibians remains intact. ... All that is required to accommodate the discovery is to add another branch to the prevailing taxonomic tree. Such a discovery can be exciting, but it would not be a revolutionary discovery in the Kuhnian sense. (Wray 2018, p. 211)

Where does this leave the notion of radical conceptual change marked by incommensurable paradigm shifts? Wray continues

Much of the apparatus associated with the earlier paradigm-related notion of revolutionary theory change was retained in Kuhn's more recently developed lexical change model of theory change. For example, he continued to believe that anomalies played a crucial role in the process of scientific change that ultimately lead to revolutionary changes of theory ... [leading] into a state of crisis ... that persistently resist[ ] resolution or normalization. The appeal to lexical changes is meant to bring into focus the essential characteristic of radical theory change .... Normal science, on the other hand, can be conducted effectively with the conceptual resources supplied by the prevailing scientific lexicon. (Wray 2018, p. 212)

But this hardly gives us necessary and sufficient conditions pinning down the notion of a Kuhnian revolution. Like the Sun, Pluto has been demoted from the status of a planet but this was hardly revolutionary. Again, in the course of the steady accumulation of knowledge in the study of hydrogen bonding, this is no longer thought of as an intermolecular force and is now regarded as a species of chemical bonding. The largely "empirical criteria" introduced in the early 1960s (Pimentel and McClellan's definition) counted the B–H–H bonding in boranes, so-called agostic interactions and interactions in halogen complexes such as HF· · ·ClF as hydrogen bonding, but recent thinking on the subject counts these interactions as not belonging to the species hydrogen bonding. On the other hand, hydrogen bonding is recognised to be far more widespread than was thought around 1930, when it was confined to fluorine, nitrogen and oxygen as donor elements and now includes carbon, phosphorus, sulphur, chlorine, selenium, bromine and iodine (see Needham 2013 for details). These changes of view arose in the course of a steadily expanding study on the part of many chemists contributing to a better understanding of hydrogen bonding since the inception of the concept in the early 1920s, during which time the concept has been gradually developed and enriched by critical reflection on earlier contributions. Wray's criteria fail to make a clear distinction between the gradual enhancement of, as distinct from the revolutionary change in, conceptual resources.

In a postscript added to the 1970 edition of *The Structure of Scientific Revolutions* Kuhn says he is not really a relativist because he is "a convinced believer in scientific progress" (p. 206). Continuing with what he apparently takes to be the central issue, he opposes the view that "A scientific theory is better than its predecessors not only in the sense that it is a better instrument for discovering and solving puzzles but also because it is somehow a better representation of what nature is really like" (loc. cit.). This comment has a distinctly realist flavour, which is apparently countered when he goes on to say that the nub of his thesis is about truth. "There is, I think, no theory-independent way to reconstruct phrases like 'really there'; the notion of a match between the ontology of a theory and its 'real' counterpart in nature now seems to me illusive in principle" (loc. cit.). Kuhn is alluding in the second part of this comment (after the semicolon) to a certain theory of the nature of truth known as the correspondence theory. There is a long and venerable tradition of philosophy in which this theory has been elaborated and defended by its protagonists and opposed by alternative views. But Kuhn barely enters the fray, and it would take us well beyond the scope of this book to pursue this topic in detail. Suffice it to say that many philosophers have questioned the correspondence theory without for a moment thinking this implies relativising the concept of truth or undermines a realist stance. Many, like Descartes, have taken the view that truth is an indefinable notion:

> I have never had any doubts about truth, because it seems a notion so transcendentally clear that nobody can be ignorant of it. ... What reason would we have for accepting anything which could teach us the nature of truth if we did not know that it was true, that is to say, if we did not know truth? ... no logical definition can be given which will help anyone to discover its nature (Letter to Mersenne, 16 Oct. 1639; Descartes 1991, p. 139).

Frege elaborated this argument that any attempt at defining the nature of truth is circular:

> a correspondence is a relation. But this goes against the use of the word 'true', which is not a relative term .... But could we not maintain that there is truth when there is correspondence in a certain respect? But which respect? For in that case what ought we to do so as to decide whether something is true? We should have to inquire whether it is true that an idea and a reality, say, correspond in the specified sense. And then we should be confronted by a question of the same kind, and the game could begin again. So the attempted explanation of truth as correspondence breaks down. And any other attempt to define truth also breaks down. For in a definition certain characteristics would have to be specified. And in application to any particular case the question would always arise whether it were true that the characteristics were present. So we would be going round in a circle (Frege 1918, pp. 3–4).

He went on to conclude,

> So it seems, then, that nothing is added to the thought by my ascribing to it the property of truth. And yet is it not a great result when the scientist after much hesitation and laborious researches can finally say 'My conjecture is true'? The meaning of the word 'true' seems to be altogether *sui generis*. May we not be dealing here with something which cannot be called a property in the ordinary sense at all? (Frege 1918, p. 6).

It is doubtful whether Kuhn's claim has any worrying implications about the possibility of rational debate on the merits of competing theories or the gradual refinement and development of concepts that we often see in the history of science.[4]

---

[4]For further discussion of the notion of truth, see Asay (2018).

As for the first part of Kuhn's comment, about theory independence, note that in defending here the possibility of genuine conflict, no appeal was made to finding theory-neutral descriptions. Lavoisier understood dephlogisticated air to be what he called oxygen. It is still often suggested that the import of the theory-laden terms can be given in terms of theory-neutral descriptions by so-called operational definitions.[5] The idea of operational definitions was introduced by the American physicist Percy Bridgman in the 1920s in the same spirit and at the same time as the logical positivists in Europe with their verification principle were suggesting that the meaning of theoretical terms be referred to theory-neutral observations. The most devastating systematic criticisms of this idea were developed by some of the former positivists themselves as they came to see the error of their ways (Hempel 1950, 1954). One of the things Bridgman stressed was that each term should have a single operational criterion of application. Even when two different operational procedures appear to agree (as, for example, optical and tactual criteria of length) there is always the possibility that future investigation will reveal a discrepancy. And different operational procedures would be necessary to measure what is usually regarded as a single concept, for example temperature, over the different ranges accessible to different kinds of thermometer. Any such identification therefore presupposes an inductive inference, bringing the generalisations of theory into the determination of the meaning of terms and thus defeating the purpose of the whole enterprise. How far should such scruples be taken? Clearly, an inductive inference is involved in assuming that every time the calx of mercury is heated, the same gas (oxygen, dephlogisticated air) is produced. So if the doctrine were to be taken seriously, a different term should be introduced for whatever gas is produced when substances looking like the red calx of mercury are heated. And of course, failing to follow the spirit of the doctrine to this absurd limit is to reject the doctrine.

Many of the positivists advocated a view remarkably similar to Kuhn's on the question of truth. They called it the instrumentalist view of theories. According to this theory, only strictly observational claims—in terms of which operational definitions were to be formulated—are true or false. Like Kuhn, they too maintained that there is no basis for claiming that the entities with which a theory populates nature are really there. In rejecting their notion of theory-independent observation, Kuhn seems to have extended this thesis to the total rejection of truth.

It has not been necessary to resort to any such doctrine of theory-independent description in order to defend the possibility of rational debate. Drawing attention to the difficulties facing this doctrine is not sufficient to motivate the thesis of conceptual and theoretical relativism advocated by Kuhn and his followers. It has been suggested that the use of foreign terms presupposing theories we reject is not an insuperable obstacle to translating what others have said with their help into our own language. This facilitates a direct conflict of opinion rather than obliging us to acquiesce in the abysmal condition of not being able to come into meaningful contact with our fellow humans.

---

[5] By Bjereld et al. (2002), for example, despite their routine disavowals of positivist ideas.

The Kuhnian rejection and replace view of historical progress has been challenged in the case of Lavoisier's contribution to the theory of chemical reaction and the development of the concept of hydrogen bonding during the last century has been put forward as an example of successive, step-wise conceptual development in the context of very extensive studies. An example of a much longer conceptual development is given in the next section with the development of Euclidean geometry over more than two millennia, emphasising the role of critical appraisal of earlier work presupposing understanding and engagement with earlier contributions. Some aspects of the Euclidean approach to systematic theorising were involved in the gradual development of the physical concept of pressure, albeit over a considerably shorter period of one century, which is taken up in Sect. 3.4.

3.3   Euclidean Methodology

The gradual development of Euclidean geometry provides an example scientific progress over a long period of time which doesn't conform to Kuhn's conception of revolutionary progress. Euclid formulated what came to be called Euclidean geometry in his *Elementa*, around 300 BC. His system was based on 5 axioms, from which all the properties of space were to be deduced as theorems with the aid of a number of general principles or "common notions". Four of his axioms could be formulated in relatively short sentences. The first two, for example, read:

1. A line can be drawn from any point to any point.
2. Any straight line can be extended continuously in a straight line.

The simplicity of these postulates was held to be a virtue, enabling the reader to see that they must be true. The famous fifth axiom was more complicated, however, requiring several lines:

5. If a straight line crosses two other straight lines so that the sum of the two interior angles on one side of it is less than two right angles, then the two straight lines, if extended far enough, cross on that same side.

For some two millennia mathematicians tried to redress what they saw as a failing in this complex fifth postulate by trying either to show that it could be replaced by another axiom as short and obviously true as the other four, or to show that it was redundant in the sense of being deducible from the other four.

At the end of the nineteenth century the German mathematician David Hilbert published a study of Euclidean geometry (Hilbert 1899) which reflected the attention paid to the standards of rigour that mathematicians had come to demand of their proofs. Hilbert criticised Euclid by claiming that many of the theorems of the *Elementa* don't in fact follow from the five axioms. Euclid's proofs were not good enough. Yet it seemed that the theorems were true—they seemed to correctly describe properties of space as we ordinarily understand it, and Hilbert certainly had no thought of rejecting the theorems. But how could the edifice be repaired? It wasn't simply a matter of adding additional lines to Euclid's proofs, or providing alternative proofs. The five axioms were inadequate to the task. Hilbert had to add additional axioms, quadrupling the number to a total of twenty axioms, in order to be able to provide complete proofs of the theorems. Moreover, several of the axioms he added were at least as complicated and foreboding as Euclid's fifth postulate. The research programme that had inspired mathematicians over two millennia came to a halt with the realisation that it was not going to be possible to summarise the whole truth about Euclidean space with a set of axioms as simple and few in number as had been hoped.

Was the increased rigour demanded of proofs a result of a change in the notion of a proof? Was the claim that such-and-such was a proof that the angle sum of a triangle is equal to two right angles true for Euclid and false for Hilbert? Had the concept of a proof changed in the course of the two and a half millennia separating Euclid's *Elementa* and Hilbert's *Grundlagen*? If the latter were the case, we would expect mathematicians to have gone to considerable lengths to formulate what they thought were the rules of proof that Euclid followed and the rules of proof that they followed, so that the differences could be clearly identified and explicitly incorporated into new work. But there is little evidence of this during the period at issue; proof theory is a twentieth century phenomenon. No. What mathematicians after Euclid were doing was criticising him. They pointed out that at different stages in his proofs, he would implicitly appeal to what he took to be true rather than rigorously falling back on his axioms. (We will see in the next section that Stevin did the same in his attempt to formulate hydrodynamics.) The whole point of an axiomatisation, they insisted, is that all and only the basic propositions to be used in a proof be articulated explicitly and included in the axioms. There can be no implicit appeal to propositions outside the system. But note that if we were to accept the relativist's interpretation of the situation, and say that whether an argument is a proof is true for Euclid and false for Hilbert, then the possibility of criticism and subsequent improvement is totally undermined. Euclid and all the mathematicians that followed him were all doing their own thing. Fantastic! Nothing anyone does encroaches on what anyone else does and everyone is happy. All these mathematicians were busily filling paper with pretty patterns, as each found it most delightful to do, and not in accordance with any general standards—an emerging set of standards satisfying humanity—because there aren't any. But if everyone does his own thing, how can steps in a proof be distinguished as reasonable or unreasonable? How is criticism possible if flaws in reasoning are not to be established by reference to universal principles?

Although the meaning of "proof" has not changed over the period at issue here, an interesting change in interpretation of one of Euclid's concepts has taken place. Familiar Euclidean notions such as point, circle, line, triangle, cube, and so forth, retained the same significance for later mathematicians as they had for Euclid. So axioms such as the first one remained as intelligible as they always had been. But what does it mean to say that a straight line can be continuously extended in a straight line, as the second axiom (see above) has it? The problem of interpretation arose in determining what the consequences of the axioms really are, and came to a head in the light of the famous "proof" of Gerolamo Saccheri (1667–1733) with which he claimed to show that the fifth postulate was deducible from the other four. The import of the fifth postulate (as Playfair showed) is that through any point outside a straight line, there passes exactly one straight line parallel to the first in the same plane. Saccheri thought he had shown by reductio ad absurdum that, in effect, the other four axioms made it impossible, either that there should be no parallel through the point, or that there should be more than one parallel through this point. In other words, if he was right, it is not possible to deny Euclid's fifth postulate given the other four. In fact, Saccheri hadn't shown that the several parallels alternative actually led to a contradiction. It led to strange consequences, but as mathematicians came to realise, this corresponded to a strange, perhaps, but perfectly consistent non-Euclidean geometry which was later systematically developed by Lobachewsky. The other alternative, the no parallels postulate, did lead to a genuine contradiction, however, and so was excluded by the other four axioms. Yet around the same time Lobachewsky developed his non-Euclidean geometry, Bolyai and Reimann developed another form of non-Euclidean geometry encompassing the no parallels postulate. And this new geometry was no less consistent than the other two systems. There seems to be a conflict, then. How was it resolved? Is there a mistake in Saccheri's proof of the contradiction? No. The issue is more subtle. There is an ambiguity in the second axiom. Saccheri's interpretation leads to the contradiction, and another interpretation allows the no parallels geometry. According to Saccheri's interpretation, continuously extending a straight line results in a straight line that is ever longer. This is the natural way to understand the wording in the context of Euclidean geometry. But the words can also be interpreted in a weaker fashion, to mean that the process of extending a straight line will never come to an end in the sense that no end point, beyond which no extension is possible, will ever be reached. This interpretation can be applied to a segment of a great circle on a sphere, which has no end point. It doesn't imply that there are infinitely long straight lines, as in the Euclidean case. Once the ambiguity was appreciated, terminology was developed which allows for a clear distinction between these two interpretations. Again, constructive criticism leading to new insights and discoveries would not be possible on the relativist's view that continuous extension for Euclid is one thing and continuous extension for modern writers is something completely different.

3.3.1 A Note on Holistic Import

We can note, finally, that there is some support in this discussion for the holistic idea that the meaning of a term or a sentence is dependent on the whole system in which it figures. It is natural to look to the consequences of a sentence, and in particular an axiom, in order to gain an insight into its meaning. We saw above Euclid's second axiom led to a contradiction when taken in conjunction with the no parallels postulate in Saccheri's argument. Consequences don't follow, however, when the sentence is considered in splendid isolation. It is only in conjunction with other principles laid down in the system—the other axioms, the definitions and the principles of logic—that consequences follow. But this doesn't mean that any change in the ancillary principles entails a radical change of meaning in the given sentence. We saw this in the case of Lobachewsky geometry, in which Euclidean parallels axiom is changed to allow that through a point outside a given line, more than one parallel line passes in the same plane. The second axiom is retained there with much the same import in this new context.

Saccheri's argument clearly involves logical principles. He sought to establish the fifth Euclidean postulate by reductio ad absurdum reasoning. This involves showing that denying the fifth postulate leads to a contradiction. Since a contradiction is false, logic (more specifically, the principle of modus tollens) dictates that what implies it, namely the denial of the fifth postulate, is false, so that it is not the case that it is not the case that the fifth postulate holds. Implicitly relying on the logical principle of double negation, he concluded that the fifth postulate holds.

Finally, definitions of terms can depend on the other principles in the system. A simple illustration is the definition of the number zero in the system of real numbers. Of the several axioms for the real numbers, two are as follows:

1. For any x and y, $x + y = y + x$.
2. There is an x such that for any y, $x + y = y$.

It follows from first of these axioms that the number said to exist in the second is unique. For suppose there is an x such that for any y, $x + y = y$ and an x' such that for any y, $x' + y = y$. Taking y in the first case to be x' and y in the second case to be x, we have

$$x + x' = x',$$

$$x' + x = x.$$

Now by the first axiom, the left hand sides of these two equations are identical, in which case so are the right hand sides. Accordingly, there is exactly one object satisfying the condition in the second axiom, sanctioning the definition of the term "0" by

Def   0 is the real number such that, for any y, $0 + y = y$.

The adequacy of this definition depends on the axioms which assert that there is something to which the singular term refers and that it is not ambiguous, referring to several objects.

3.4 Pressure: An Example of Progress in the Articulation of a Concept

Isaac Newton (1643–1727) famously declared in a letter to Robert Hooke in 1676, "If I have seen further it is by standing on the shoulders of Giants", reflecting the fact that many of his outstanding achievements in physics can be seen as a continuation and (often considerable) extension of lines of thought on the part of his predecessors. What clearer declaration could there be from one of the most important figures in the history of science opposing the Kuhnian notion of progress by gestalt shift from incommensurable precursors?

One field in which Newton saw further is hydrodynamics. His formulation of the basic principles depended upon the development of the concept of pressure. This is a tale of steady scientific progress by building on past gains and critical appreciation of past deficiencies. It was as much a matter of coming to see and explicitly articulate implicit understanding of the workings of nature as the discovery of new principles and provides a good example of how science progresses. I shall be drawing heavily on the recent monograph by Alan Chalmers (2017), which I strongly recommend the interested reader to consult for further details.

Chalmers recounts the development of the concept of pressure during the century following the publication in 1586 of *Elements of Hydrostatics* by Simon Stevin (1548–1620) to Newton's *Principia* in 1687. We take for granted the modern concept of pressure, represented by the "P" in the gas law $PV = nRT$, as a scalar magnitude of an intensive quality (every spatial part of a body exerts the same pressure as does the whole). But the first thinkers who contributed to the development of this notion thought of pressure as a directed magnitude. Indeed, the modern definition of pressure as force per unit area might be taken to suggest as much. This is something of a paradox, suggesting that the vector notion of force is given as the product $P.A$ of pressure times area, but which isn't a vector![6] The story of the development of the concept of pressure is partly the unravelling of apparent paradoxes, in the course of which we can see something of the emergence of the modern empirical view of scientific knowledge.

Stevin set out his theoretical treatment on the basis of principles he took to be obviously true that could be taken as given at the outset, and which he envisaged as forming the basis of an axiomatic system modelled on Euclidean geometry. The role of observation and experiment, on his understanding, was to illustrate these preconceived ideas, not to provide evidence for them or reveal novelties. Unexpected phenomena would be shown to be distant consequences of familiar principles by rigorous deduction along the lines of geometric proofs. Stevin sought to extend the extant science of simple machines to include hydrostatics and looked to the ancients for guidance not only in methodology but also regarding his specific subject matter. This he found in Archimedes' treatment of floating, which assumed the weight of water above a horizontal surface is the only force acting on it (leaving it as a mystery how the downward thrust of displaced liquid could turn 180° and be transformed into an upward thrust on the immersed solid). The idea of weight as the fundamental driving force is one Boyle, following Galileo, sought to dispel by demonstrating that a solid body could float in a minuscule amount of water weighing much less than the solid body. Stevin also failed to explicitly acknowledge that the liquid state of water, which is what he had to deal with, was crucially different from

---

[6]In modern mathematical notation, an area has a directed vector quality of orientation, specified by a unit vector **i** normal to the surface, so that the force on a surface can be written $\mathbf{F} = P.A.\mathbf{i}$.

the nature of solid matter with which arguments deriving from Archimedes were concerned. This was rectified in degrees by Blaise Pascal (1623–1662), then Robert Boyle and finally Isaac Newton, who articulated the appropriate notion of the liquid state as exhibited by matter every part of which is unable to resist distorting forces in any direction. Yet Stevin was a successful engineer who implicitly accommodated the difference between solid and liquid in his engineering practice. For example, it was he who designed lock gates to shut in a "V"-formation pointing upstream, causing the water pressure to hold the gates tightly sealed. The water is acting horizontally on the gates, but the horizontal component of the water's weight is zero. In fact, Chalmers maintains that all the empirical evidence that was eventually called upon in support of the theory of hydrodynamics that finally crystallised at the hands of Newton a century later was essentially already at play at the time Stevin was practicing.

Stevin's empirical knowledge was not incorporated into his postulates. Some was formulated in his theoretical treatment, deriving, presumably, from his experience as an engineer, which did deal correctly with hydrostatic forces on planes other than horizontal and implied that hydrostatic forces bend around corners. But it rested on arguments that were invalid and did not, in fact, follow from his postulates. It was in effect introduced as independent material. There was a common-sense notion of pressure at work here, involving a liquid pressing against a solid surface, but this is not the technical notion of pressure needed to explain how it is transferred from one place to a distant one within the same body of liquid. It was Newton who finally succeeded in articulating an appropriate definition of liquid and proving theorems which did follow from his explicitly stated principles in accordance with the Euclidean ideal and which did accommodate the empirical knowledge implicit in Stevin's practice. Leonhard Euler (1707–1783) later tidied up the mathematical formulation of hydrodynamics by exploiting mathematics of the continuum not available to Newton and generalised the theory to include compressible fluids, i.e. gases, so incorporating what Boyle's experimentation in pneumatics led him to characterise as the "spring" of air, distinct from its weight. His recognition of the spring of air as a property of every part of a given body of air with no analogue in solid bodies must, Chalmers suggests, have alerted him to how liquids, with no significant spring, nevertheless differ from solids. Building on Pascal's work, he exploited the device of considering forces acting on either side of imaginary planes within the body of a liquid at equilibrium to show that the force per unit area on such a plane depends only on its depth, and not at all on its orientation, which was a step in the direction of dissociating pressure from weight. Exploiting Boyle's device, Newton considered the pressure on arbitrary planes within a sphere of incompressible liquid devoid of weight pressed equally over its surface. He recognised the necessity of assuming that fluids are completely continuous, *all* their parts being unable to resist distorting forces, in order to show how, unlike solids and powders,[7] fluids transform directed forces into isotropic ones at equilibrium, with

---

[7] Water, unlike a powder, poured into one arm of a U-tube comes to equilibrium when the water in both arms reaches the same height. This remains true if one arm is of considerably larger cross-section, when the weights of water in the two arms that balance one another are unequal.

pressure acting equally in all directions. This clearly distinguishes pressure from weight, which could then be put in as a separate and additional factor affecting the behaviour of fluids.

Newton's thesis of the continuity of fluid matter is at odds with ideas Descartes put forward in response to his criticism of what he saw as Stevin's failure to provide a physical account explaining his hydrostatic claims. Like many of his contemporaries, Descartes sought a mechanical explanation of the behaviour of matter in terms of how corpuscles impinge on one another whose weight derives from centrifugal forces generated by vortices. Mechanical explanations of this kind, relying ultimately on size, shape and motion of the ultimate particles, were held to be intelligible, unlike obscure theories heralding from medieval scholastics. But such a priori reasoning faulted even when it came to providing deeper explanations of the workings of simple machines such as the arm balance, which relied on properties such as rigidity, weight and elasticity that were not successfully reduced to the properties of shape and motion that proponents of the mechanical philosophy allowed their corpuscles. Boyle was astute enough to realise that no known corpuscular theory was adequate to this task, and although he never abandoned his faith in ultimate mechanical explanations, he was content to account for his experimental results in terms of "intermediate" causes such as the spring of air, the weight of bodies, the rigidity of solids and the fluidity and pressure of liquids. It was very clear to Newton that Descartes' corpuscular theories of matter were unable to explain the transmission of pressure in fluids, and he didn't dabble in preconceived hypotheses with no demonstrable application. The "experimental philosophy" that Newton endorsed in the spirit of Pascal's dictum that "experiment is the true master that one must follow in physics" (Chalmers 2017, p. 176) didn't begin with preconceived principles. Pascal justified his somewhat counterintuitive principle, that a force applied at one location on the solid surface bounding a liquid will appear as the same force per unit area at any other location on the bounding surface, by showing how a wide range of hydrostatic phenomena could be understood as consequences of this principle. If the principle was antecedently problematic from the point of view of common sense or the mechanical philosophy, so much the worse for them. Newton established the adequacy of his definition of a liquid in the same way, and this is how he went on to defend his law of gravitation despite its involving what many thought to be the counterintuitive idea of action at a distance (to which I return in Sect. 7.2).

3.5 Taking Stock

Relativism has ancient roots which have sprouted new adherents throughout the ages, notwithstanding the classical dilemma. The relativist may give vent to his opinion but is powerless to defend it when challenged. For if no sincere claim which seems to the beholder to be true can be judged false by appeal to some considerations raised by anyone else, but merely counts as false for the antagonist, then the distinction between truth and falsehood has been abandoned. Whatever else it might be, then, the relativist cannot consistently maintain his doctrine is true. Without recourse to some recognised standards, the relativist cannot offer any justification or rebut the antagonist's challenge by showing it unjustified. We might have thought that this ancient argument closed the issue once and for all. But the essential strands of this exchange have resurfaced in recent times and the ancient wheels continue to turn.

Kuhn raised the spectre of relativism in the second half of the twentieth century with his doctrine of paradigm shifts driving revolutionary changes of scientific theory and introducing new claims incommensurable with those they replace. These were exciting new ideas when they first appeared in the early 1960s, which challenged the remnants of logical positivism enshrined in what was known as the received view of scientific theories—in particular, the positivist's distinction between observational and theoretical terms. This distinction had already been questioned by Quine in a widely-read article published in the early 1950s on the strength of an argument essentially the same as that Pierre Duhem had put forward at the beginning of the century. But whereas Quine was a logician who put the argument in abstract terms that caused a stir amongst philosophers of language, Kuhn was addressing science directly and made a more forceful impact on the philosophy of science. In fact, a lasting feature of work in the philosophy of science from this time is the concern with the history of science, both recent and more distant, for which Kuhn can in large part take credit.

Kuhn's ideas about revolutionary change made a more lasting impact in the social sciences than on philosophers, who were more sceptical of relativist doctrine. He did, as we saw, try to temper this aspect of his thought in a later edition of *The Structure of Scientific Revolutions*. But the import of his comments on the notion of truth were at best unclear. Attempts at sharpening the terminology haven't significantly improved on the unclarity and certainly seem to take the sting out of what made his ideas attractive in the first place. Characteristic Kuhnian terminology still makes sporadic appearances in the literature, but with non-revolutionary import. Chang (2012a), for example, describes the concepts of Brønsted-Lowry acid and Lewis acid as "incommensurable". This merely dramatises what was described above as two different ways of generalising the traditional concept of an acid and involves no gestalt shift raising a barrier to comparing the two, as is eminently clear from Chang's account. Wray (2018) claims that the move from basing the periodic table on atomic number rather than atomic weight marks a scientific revolution. But his discussion shows nothing of the probing of conceptual depths displayed in Chalmer's discussion of the *development* of the concept of pressure. During the decades of the nineteenth century when the periodic correlation of chemical properties with atomic weight was first suggested, there was no notion of what an atom was and atomic weight was merely a number without units expressing a proportion. It is simply correlated with a particular elemental substance (i.e., a substance, as Lavoisier defined "element", which cannot be decomposed into other substances). So it would be misleading to describe the change as a switch from atomic weight to atomic number with all the import that these terms have acquired in the twentieth century, in terms of electrons and nuclear constituents. It was already becoming clear in the latter part of the nineteenth century that the application of classical ideas to the atomic realm was paradoxical (Maxwell 1875 [1890]) and something radically new was required. Many puzzles were resolved and novel properties investigated when the first empirically justified ideas about what atoms are emerged at the beginning of the twentieth century and were developed from the old quantum theory to the wave mechanics of the 1920s and beyond. This era in the history of science is generally described as the quantum revolution, of which the recognition of atomic number was just one part. (Developments in the sixteenth century comprise what is generally described as the other scientific revolution, of which the development of the concept of pressure within hydrodynamics was a part.) The quantum revolution consists in the *development* of a radically new theory rather than the mere contrast between old and new. Kuhn himself made a notable contribution to the understanding of this development with his *Black-Body Theory and the Quantum Discontinuity 1894–1912* published in 1978. Interestingly, in this book which, according to the fly-leaf, "traces the gradual emergence, during the first decade of the present [i.e., twentieth] century, of the concept of discontinuous physics", he makes no mention of paradigms, incommensurability or anything suggestive of his radical notion of scientific revolutions. On the contrary, he is at pains to show how Planck's theory of discrete packages of radiation emerged from an earlier interest in thermodynamics by successive reformulation and modification of his theory of radiation, during which he resisted quantising radiation as long as he could, but was ultimately faced with accommodating new empirical results.

It has become increasingly difficult to see how a latter-day Kuhnian version of scientific progress (of which we saw he is "a convinced believer") differs significantly from the thesis of the more or less gradual and essentially continuous development of science advocated by Pierre Duhem (1861–1916). It was argued that Lavoisier's new chemistry illustrates the successive refinement of old theory and emergence of new by rational dialectic rather than a Kuhnian shift to a new theory incommensurable with the old, and this conception of progress was further exemplified here in Sects. 3.3 and 3.4 with Euclidean geometry and hydrodynamics as well as the brief mention of hydrogen bonding in Sect. 3.2. We will see more of Duhem's thesis in Chaps. 6 and 7.

Chapter 4
The Use and Abuse of Science

4.1 Introduction: The Misuse of Science

The claims of objectivity mean, as we have seen, that gaining knowledge is an essentially social phenomenon, pursued with a view to satisfying the demands of public scrutiny. And like other social phenomena, there is a moral dimension to this activity. More spectacular aspects of this are familiar enough. Scientists accused of having created weapons of mass destruction have sometimes rebuffed criticism with the claim that it is politicians, and not they themselves, who have deployed the fruits of their labours. Other scientists have felt that responsibility cannot be circumvented by passing the buck in this way, even if they are not alone in bearing the burden. Many of those involved are protected from moral condemnation by a shroud of secrecy.

Where science is conducted in the open, scientists are interested in receiving recognition for their work, and sometimes remuneration. Such is the importance of priority that scientists are sometimes prepared to resort to theft. In a recent case, Robert Gallo at the National Institute of Health in U.S.A. claimed to have independently isolated a strain of the HIV virus completely different from that first isolated by the Institut Pasteur in Paris. In fact, he cultured the French virus which had been sent to him from Paris. Gallo's claims were subsequently shown false by direct DNA analysis, and the cover-up, aided by the some agents of the U.S. government, blown. It seems glory was not the motive in this case; large amounts of money were at stake in connection with a patent for a blood test based on the alleged isolation of the virus (Lang 1998).

In the early 1930s, E. Rupp published a series of papers reporting results of certain scattering experiments in agreement with theoretical calculations of Nevill Mott which brought into question Dirac's theory of the electron. The results were retracted in a note Rupp published in 1935, however, in which a doctor's report is quoted to the effect that Rupp had unknowingly suffered from a "mental weakness" (psychasthenia) under the influence of which dreamlike states where allowed to

© Springer Nature Switzerland AG 2020
P. Needham, *Getting to Know the World Scientifically*, Synthese Library 423,
https://doi.org/10.1007/978-3-030-40216-7_4

intrude into his research. Louis de Broglie comments "Rupp confessed he had made up his experiments and a little later he went mad. Therefore the situation is not clear, and his experiments could not be repeated, which proves they are falsified". Various anecdotes are in circulation, one claiming that when his locked laboratory door was opened, no apparatus for electron scattering was found (Franklin 1986, p. 228). Subsequent theoretical work in the 1940s showed that Rupp's results could not possibly have been obtained under the conditions he described. Rupp's intentions in reporting the fraudulent results are unknown. Whether it was entirely due to his mental illness, or deliberately fabricated, cannot be determined. Although his results were anomalies for Dirac's theory, this was corroborated by other experiments and therefore not abandoned. Experiments connected with the anomaly were repeated many times until the discrepancy was finally resolved.

Even Galileo—hero of the scientific revolution—was not above doctoring the results of his investigations to suit his purpose. His treatment of pendulums is a case in point. He exploited the pendulum's supposed isochronism—that for a given length, the period of a complete swing is independent of the material of the bob. The fact that bodies differing in density acquire the same speed when falling freely through the same height was demonstrated, he maintained, by letting the fall be repeated many times and accumulating these distances by allowing two bodies, one lead and the other cork, suspended by fine threads of equal length, to swing freely after being released simultaneously. "This free vibration repeated a hundred times showed clearly that the heavy body maintains so nearly the period of the light body that neither in a hundred swings nor even in a thousand will the former anticipate the latter by as much as a single moment, so perfectly do they keep step" (Galileo 1638, pp. 84–5). Unfortunately, such claims "bear little relation to actual observation, as anyone who has carried out these experiments is well aware", says Naylor (1974, p. 34), who had "yet to see the cork pendulum that will complete a thousand oscillations—indeed, only the most massive cork bobs are capable of completing even one hundred oscillations" (p. 35).[1]

There are motives between the extremes of greed and insanity, driving investigators to misrepresent their findings and mislead their listeners, falling on a continuous scale from culpability to innocence. Blame for misuse does not always lie with the researcher; journalists and popularisers misrepresent in poorly considered attempts at simplification; politicians and others seeking to influence public opinion misuse statistics, conceal uncertainties or raise unreasonable standards of certainty as

---

[1]But a more general claim by Koyré that it would not have been possible to carry out any of the experiments and observations Galileo reported goes too far. Settle (1961) repeated an experiment on inclined planes in accordance with Galileo's description, which Koyré had described as completely worthless, and found the ingenious device for measuring time gave quite precise results—certainly precise enough to attain the relations of proportion between distance and times that Galileo claimed. Koyré maintained further that Galileo's procedure couldn't possibly furnish a reasonable value of the constant of proportionality appearing in the algebraic expression of this relation of proportion. But as Settle points out, this modern way of expressing the law of free fall by writing distance as a function of time was not the way Galileo expressed the relation, which was weaker and didn't entail all that the modern functional expression does.

suits the occasion. Responsibility for the correct representation and application of knowledge on the part of agents outside the narrow circle of specialists and investigators is taken up later on in this chapter. For the time being attention is focused on the practitioners themselves and the errors of their ways.

The phenomenon of fraud is by no means new. Plagiarism is the classic form of fraud. When Newton finally published his theory of gravitation in 1687 Robert Hooke accused him of plagiarism (although without justification; see Westfall (1980, pp. 386–7)). Who knows what misdemeanours were perpetrated in the ancient world, which has left so few documents to posterity? A somewhat older contemporary of Newton's, Robert Boyle, is sometimes lauded for his promotion of atomism in works such as *The Sceptical Chymist*, which supposedly advanced chemistry by providing a physicist's theoretical interpretation of the facts provided by technically competent but theoretically incoherent chemists. This assessment is thrown into doubt, however, by investigations which show that the experimental evidence he adduces as well as the interpretation he puts on it is taken directly from a tradition of chemists reaching back through the renaissance and medieval times with roots in Aristotle, particularly as formulated in the works of Daniel Sennert (1572–1637), medical professor at Wittenberg (Newman 1996). Perhaps we shouldn't be too quick to judge Boyle for plagiarising Sennert, however, because it was a widespread practice in the English renaissance, even if writers were increasingly stigmatising the practice (White 1938). It is doubtful whether such sympathy should be extended to Lavoisier for neglecting to acknowledge his debt to Priestley, who visited Paris whilst accompanying his patron William Petty, the second Earl of Shelburne in 1774. Priestley described his new experiments with gases to Lavoisier, who eventually succeeded in reproducing them. The modern view, however, definitely condemns the practice, and seeks to redress stealing by using the work of others without acknowledgment and assign credit where it is due. Grave charges must be justly evaluated, but even where culpability is not the main issue, the truth must out, and it is important that the organisation of science makes provision for this.

4.2 Abusing the Right to Declare "I Know"

In his 1830 book *Reflections on the Decline of Science in England*, Charles Babbage distinguished three categories of fraud in science: forging, cooking and trimming. Babbage's taxonomy of fraud suggests a scale of decreasing blameworthiness. Trimming, he explained, "is perhaps not so injurious … as cooking. The reason for this is, that the average given by the observations of the trimmer is the same. … His object is to gain a reputation for extreme accuracy in making observations" (Babbage 1830, p. 178).[2] The failing of unjustly seeking acclaim for prowess as

---

[2]Note that in modern usage introduced towards the end of Sect. 2.2, we should say "precision" rather than "accuracy".

a good experimentalist seems to have applied to Millikan's determination of the charge on the electron.

In his famous 1913 article, Millikan declared that once he began making measurements, a total of 58 drops constituted all his data, with the words "this is not a selected group of drops but represents all the drops experimented upon during 60 consecutive days" (Millikan 1913, p. 138; emphasised in original). He wanted to assure his readers that he wasn't selecting from his data in unreasonable fashion. Leaving aside initial observations in the course of adjusting his apparatus was fair enough. But once he was happy with his procedure, he claimed that he hadn't left out of account any drops which might have brought into question the whole idea of a fixed minimum charge or, less drastically, lay further from his final average and increased the statistical spread on which his estimated margin of error is based.

His laboratory notebooks contradict this, however, and indicate that during the period concerned he in fact collected data from 175 drops. Even counting from the first published drop, there are still 49 drops not reported. It is reasonable that Millikan threw out earlier data when he was still unsure whether his apparatus worked as it should. But why did he continue to select from his later data? The unpalatable conclusion seems to be that he trimmed his data in order to reduce his experimental error and give a misleading impression of exactly how precise his results were.

Millikan is criticised for having been mistaken in the value of the viscosity of air, resulting in an error of somewhat less than 1%. There is no question of fraud in this case. No one suggests that there was any deliberate attempt to mislead. On the contrary, by reporting all the procedures motivated by his interpretation of the experimental situation, he laid himself open to criticism in such a way that his account could be put to constructive use in correcting and improving his results. Millikan's false claim about the 58 drops from which his published data was drawn constituting his entire set of data is another kettle of fish. Here it does seem that Millikan is guilty of the lesser of Babbage's trilogy categorising fraudulent behaviour.

It is important to be clear about the distinction between experimental or random error, the estimate of which is at issue in condemning Millikan, and systematic error, for which he is criticised but not morally condemned. There is a general problem of knowing when sufficient precautions have been taken to cope with the systematic error. Delicate experimental arrangements from which high precision is expected are subject to many instrumental difficulties leading initially to results far from expected values. Consistency upon repetition is not enough; the results should not deviate from accepted values. As one physicist has put it,

> ... the experimenter searches for the source or sources of such errors, and continues to search until he gets a result close to the accepted value. Then he stops! But it is quite possible that he has still overlooked some source of error that was present also in previous work. In this way one can account for the close agreement of several different results and also for the possibility that all of them are in error by an unexpectedly large amount. (Quoted from R. Birge, reporting a view of E. O. Lawrence, by Franklin 1986, p. 236).

Two important issues are raised here. First, there is an evident danger of objectivity giving way to a bandwagon effect. Is agreement with previous experiments an adequate criterion of the correctness of a result? Clearly, the problem is in some sense unavoidable because, in the absence of any general criterion of completeness that can be imposed on the search for systematic error, investigators must fall back on whatever indications they can find. A safeguard can only be provided by a willingness on the part of the scientific community to treat seriously anomalous results deviating from the accepted value. The referees and editors of journals must be sufficiently open-minded to countenance the possibility of the unexpected, the ingrained scepticism of the community at large notwithstanding, while still maintaining reasonable standards of scrutiny and in the hope that the issue will be resolved in the long run.

Second, there is a danger of investigators being driven to the opposite extreme of Millikan's trimming and overestimating error in order to avoid conflict with established views. As a device for covering the possibility of unknown systematic error, this confuses two quite distinct categories of possible error. Systematic errors are precisely not statistical, and enlarging the estimate of statistical error by what must be an arbitrary factor would simply destroy the statistical significance of the estimated error. Far from being a laudable virtue, erring on the side of safety in this way runs the greater risk of obscuring information which might otherwise lead future researchers to uncover some important fact, such as a hitherto unappreciated source of systematic error. The individual investigator might find in this device protection from criticism. But such deliberate dumping of information is self-serving, and certainly does not provide any kind of safeguard for the scientific community. On the contrary, the concealment of the truth by bloating the numerical estimate of error in this way is a deception of the scientific community, and by extension, the general public who foot the bill for the research. So far from expressing a virtue of caution, it is every bit as fraudulent as trimming, and just as reprehensible as Millikan's juggling with his figures. There is no getting away from the fact that what is held to be knowledge at one time, made as precise as possible by an estimate of the likely error in the light of the best assessment of the situation at that time, may have to be revised at some later time. An investigator who claims to know takes upon himself the responsibility of having adequate support for the appropriately precisified claim and must be prepared to confront a critical public. Timidly seeking to avoid scrutiny by reducing the claim is like the philosophers who, by avoiding the risk of being wrong, allowed the sceptics to drive them to solipsism. There is no virtue in weakening a claim with the sole motive of avoiding being shown wrong.

4.3 Fraud and Controversy

Controversy is the life-blood of science. A regime in which publications were routinely expected to go uncriticised would not be able to maintain reasonable standards of objectivity. Fraud is exposed by critical examination, and is left completely untouched by edicts from experts simply declaring in effect that the results are not what they would expect. Sometimes we might think that astrologers, mediums and churchmen make fraudulent claims, but that these are harmless enough and needn't worry us unduly. At all events, they are better left alone than countered in the same proselytising tone. But sometimes there may be good reason to expose claims which have been used to exploit a gullible public and extract ill-gained profit or influence. A case in point is the attempt to provide evidence for the underlying idea of homeophathic medicine reported in *Nature* some years ago.

Nature is a highly respected journal covering the broad range of natural science. It is more difficult to get articles accepted in the generalist journals, and so publication in such a journal as *Nature* carries high status. This, of course, is something the editorial committee of *Nature* was aware of when they accepted for publication the claims of Benveniste and his co-workers to have evidence supporting contentions in homeopathic medicine about the ability of water to preserve an influence from dissolved substances when the concentration is dramatically reduced (Davenas et al. 1988). Such is the degree of dilution that there aren't enough molecules of the original agent to go around all of the samples at the end of the dilution process. Nevertheless, it was claimed that the results of the investigation show, on a statistical basis, that the medicinal action of the agent is preserved— a result which the investigators interpreted in terms of an ability of water to take up an "imprint" of the agent which is retained throughout the dilution process. The editors and referees were sceptical, and the published article was preceded by an editorial statement under the title "When to believe the unbelievable?". An "Editorial reservation" printed on the last page of the article makes it clear that publication was accepted on the condition that Benveniste's team allowed a group from *Nature* to visit their laboratory in Paris and form their own opinion of how the experiment is conducted. The results were published in *Nature* a month later under the title "'High-dilution' experiments a delusion", where the authors say

> We believe that experimental data have been uncritically assessed and their imperfections inadequately reported. We believe that the laboratory has fostered and then cherished a delusion about the interpretation of its data. (Maddox et al. 1988, p. 287)

The central point was that the positive results failed to materialise in a double-blind experiment where samples are coded in order to eliminate observational bias by hiding from the observer whether what is being tested is a result of the diluting process or a control. Observations which bring into question principles based on centuries of observations (such as the law of mass action) and lack alternative explanation call for particularly stringent procedures. Benveniste (1988) replied, giving some sense of the atmosphere created by the inspection squad ("the hysteria was such that Maddox and I had to ask Stewart not to scream"), and complaining of "Salem witchhunts and McCarthy-like prosecutions".

In the ensuing discussion following the publication of Benveniste's results, the decision to publish was challenged by Metzger and Dreskin (1988, p. 375), who were themselves in no doubt about the matter: "We think not". The editors responded (p. 367) by defending their action and making the point that

> there is no absolute rule that journals such as this, which are proud to publish a great deal of original science, should not on occasion give an airing to material that is different in kind. ...[This] may be a public service.

They refer to the case of scotophobin, a protein supposed to record habits acquired by trained rats which could be injected into untrained rats who would thereby acquire the original rats' learned capacity to run a maze. *Nature* published a version of this story together with a devastating critique (238 (1972), pp. 198–210) which seems to have extinguished that particular line of research. "Not that belief in the magical properties of attenuated solutions will be as quickly exorcised", they regretfully add, noting that a quarter of French doctors prescribe homoeopathic medicines. But only by bringing the issue into a suitable public forum can it be competently debated.

Self-censorship and decisions by referees and editors might be expected to restrict publication of anomalous results deviating from previous observations. The fact that journals have published articles like these indicates a certain openness on the part of referees and editors. An interesting comparison might be made with another article published in *Nature*. Astronomers today are largely agreed that the universe began with a big bang. Previously, the uniform density of matter in all directions despite the continual expansion of the universe was understood by a substantial group of astronomers to speak in favour of a steady state theory. According to this theory, hydrogen atoms come into being at a rate sufficient to maintain the overall density of matter given the expansion. But the discovery of the so-called background radiation in the 1960s was deemed speak against the steady state theory and in favour of the big bang theory. This hasn't entirely quenched the discussion, however, and advocates of the steady-state theory point to anomalies which they think speak in favour of their theory (Arp et al. 1990). These authors charge defenders of the big bang theory with "avoiding confrontation with observation" which "is not the hallmark of a good theory" (p. 809), especially since

> Cosmology is unique in science in that it is a very large intellectual edifice based on very few facts. The strong tendency is to replace a need for more facts by conformity, which is accorded the dubious role of supplying the element of certainty in people's minds that properly should only belong to science with far more extensive observational support. When new facts do come along, as we believe to be the case with anomalous redshifts, it is a serious misprision to ignore what is new on the grounds that the data do not fit established conformity. Certainty in science cannot be forthcoming from minimal positions such as those which currently exist in cosmology. (Arp et al. 1990, p. 812)

There is no question here of *Nature* trying to bring into the open fraudulent negligence in the conducting of observations. On the contrary, it is a (vain?) attempt to instigate the proper procedure of criticism. H. C. Arp, the principal author of the paper in question, was obliged to move to the Max-Planck Institut für Astrophysik after being prevented from making telescopic observations at the

Palomar Observatories at Mount Wilson in the mid-1980s because his results were unwelcome. Another author, Fred Hoyle, has since died, giving some support to the unflattering generalisation of Clerk Maxwell's claim that there are two theories of the nature of light, the corpuscle theory and the wave theory; we used to believe in the corpuscle theory; now we believe in the wave theory because all those who believed in the corpuscle theory have died (quoted by Klotz 1980, p. 130).

4.4 Cooking

Babbage's classification allows for a more serious infringement of good scientific practice than trimming without going so far as to involve a wholesale forging of data. What he called cooking involves the corruption of data by reworking results in dubious fashion and presenting them as fair and correct reports of observations. A famous example is provided by the once-lauded verification of Einstein's general theory of relativity by Sir Arthur Eddington's observations of the solar eclipse in 1919, which was quoted in many textbooks in the succeeding decades.

Einstein's general theory of relativity predicts the bending of light rays in the vicinity of large masses such as that of the sun. Newton's theory of gravitation, which Einstein's theory replaces, does too, but for somewhat different reasons, and of a magnitude of just half of that predicted by Einstein's theory. The bending could be observed as the apparent shift in position of a star if observed close to the sun's surface, when the light from the distant star passes close to the sun before reaching the observer on earth. Unfortunately, such stars cannot ordinarily be seen when so close to the sun because their light is completely drowned by the intense light from the sun. But the sun's blinding light is shut off for an observer on the earth during a solar eclipse, which provides the circumstances when the observations are, in principle, feasible. Opportunities for making such observations don't present themselves very often, however. But a solar eclipse, observable in the southern hemisphere, was due shortly after the First World War. The matter was considered so important that two expeditions set out in March, 1918, one for Sobral in Brazil and the other, with Eddington, for Principe, an island off the west coast of Africa. Eddington presented the results to a well-attended gathering of the scientific elite at a meeting of the Royal Society in London on 6th November, 1919. The following day, the news was heralded in *The Times* with the headline: "Revolution in science— new theory of the universe—Newton's ideas rejected". Eddington claimed that the results were clearly in agreement with Einstein's theory and contradicted Newton's.

The problem is that the results Eddington actually obtained didn't speak so clearly in favour of Einstein's theory (Earman and Glymour 1980; a briefer summary is given in Collins and Pinch 1998, pp. 43–52). Only after a thorough "reworking" and selection of the data was Eddington able to present figures with the import he claimed. The basic strategy involved comparing photographic plates taken during the eclipse with photographs of the same region of the sky when the sun is absent. The photographs with the sun in the region of the sky in question were taken during the day, and the comparison photographs were taken at night, some several months later. Difference of season and the time of the day when exposures were taken introduced many variations of circumstances which had to be taken into account in the calculations. The difference in temperature, to mention just one factor, introduced a change in focal length itself affecting the apparent position of the stars by an amount comparable to the size of the effect that was to be measured.

Calculation of the predictions from theory was a complicated and not at all straightforward procedure, but it was taken that Newton's theory predicted a shift of 0.8 s of arc for stars very close to the sun, and Einstein's theory 1.7 s of arc, a second being 1/3600 of a degree. It was only possible to observe stars no closer than two solar diameters from the edge, however, reducing the difference to only half this. Measurement of such a small difference would ideally be made with large telescopes, which gather large amounts of light and are controlled by finely engineered mounts and machinery holding the telescope steady and moving it smoothly to compensate for the rotation of the earth. But this was not possible in the remote regions of the earth where there was a full solar eclipse. Smaller telescopes, requiring longer exposure times during which enough light could be gathered to produce clear images, with cruder mechanisms for following the earth's rotation, had to be used, introducing many more sources of error. The Sobral group took two telescopes, the Principe group one. It was cloudy over Principe on the day of the eclipse, and of 16 plates taken, only two, each showing five stars, were usable. The Sobral group obtained 8 from one of their telescopes, leading to a deflection calculated to lie between 1.86 and 2.1 s, in agreement with Einstein's theory, but 18 usable plates (though of poor quality) from the other leading to a deflection of 0.84 in agreement with Newton's theory. The two Principe plates were of even poorer quality. Yet Eddington obtained results lying between 1.31 and 1.91, albeit by a complex argument using questionable assumptions. In his presentation to the Royal Society, he maintained that the results definitely supported Einstein's theory against Newton's. But his reasoning was not so clear. As W. Campbell, an American commentator, put it in 1923,

> Professor Eddington was inclined to assign considerable weight to the African determination, but, as the few images on his small number of astrographic plates were not so good as those on the astrographic plates secured in Brazil, and the results from the latter were given almost negligible weight, the logic of the situation does not seem entirely clear (quoted by Earman and Glymour 1980, p. 78).

Eddington claimed he could ignore the Sobral results because they suffered from systematic error. But all the results were subject to many substantial sources of error, and he didn't say why what he said of the Sobral results shouldn't apply to the Principe results. The scientific authorities moved in to support Eddington, however, who was allowed to write the standard accounts of the expeditions and their interpretation.

The story as Eddington told it conforms to the simplistic popular image of science as progressing by throwing up theories making novel predictions, and allowing the conducting of crucial experiments or making crucial observations facilitating a decision between competing theories to be made on the basis of differing predictions that they make. Stephen Brush (1989), who has some not very complementary things to say about Eddington's motives, suggests that the reason general relativity was accepted by scientists is not that it predicted new phenomena that were triumphantly confirmed, but that it solved an outstanding problem for Newton's theory of gravitation, namely the anomalous perihelion of Mercury. It was known since the early nineteenth century that the observed orbit of Mercury doesn't comply with Newton's theory. A similar anomaly in Uranus' orbit led to the prediction of a not previously observed planet further from the sun which disturbed Uranus' orbit and the spectacular discovery of the planet Neptune. The same trick with a not previously observed planet causing anomalies in Mercury's orbit and optimistically named "Vulcan" didn't work, and the problem remained until Einstein's general theory of relativity offered a solution. But this shouldn't detract from the fact that the theory was controversial in the years immediately following publication. Although general relativity is now recognised as the height of Einstein's many contributions to science, when awarding him the Nobel prize in 1922 the committee was unwilling to award him the prize for this theory, and motivated it instead on the basis of his 1905 theory of the photoelectric effect.

4.5 Masquerading as Science

Whatever the pedagogical virtues of presenting major advances in science on the basis of straightforward and unambiguous experimental confirmation, there is a price to pay for such oversimplification. Any crackpot scheme can be formulated so as to conform to a sufficiently mindless and simple specification of methodological principles. Beliefs whose motivation relies on sources other than the reasoned appeal to empirical evidence and systematic interpretation are easily cast in the form of hypotheses putatively explaining the phenomena with which science tries to come to grips. Lobbying political institutions proud of their liberal traditions has led uncertain and uneducated politicians to cave in to the apparently reasonable argument that all views should be given an airing in the education of our youth. The education of the next generation plays a dominant role in forming their general outlook and equipping them with complex and not easily mastered skills with which to build their futures. Given the severe limitations on the time and energy available, the idea that effort should be spent on every view on whatever subject is surely a non-starter. What is allowed into the curriculum, depriving time and energy which could be spent on important and difficult study, should satisfy reasonably stiff criteria. Open-mindedness demands, not that we abandon our youth to the line of least resistance and apathetically grant equal status to all contenders, but use our intellectual faculties in examining the credentials of ideas espoused by others and decide whether they meet appropriate standards. Science shoots itself in its own foot by presenting simplified popular accounts clearly suggesting standards which it wants to claim in another public forum are not adequate to meet criteria for inclusion in the curriculum.

Intellectual tolerance is rooted in the quest for the truth. It recognises that once cherished ideas have proved to be untenable, and acknowledges the fallibility of modern scientific theory. Serious criticism must be taken seriously. But once it is clear that an idea makes no contribution to our understanding, it has no further claim on our tolerance to be given further attention. Rather than meeting such a challenge, creationists, for example, and revisionist historians who defend the thesis that there was no systematic extermination of the Jews by the Nazis, make their claim to our attention on other grounds. It is a truism that the ideas that we do support because well-confirmed have not been literally proved, and cannot literally be said to be absolute certainties. If popular science wasn't presented in quite the dogmatic way that it often is, parading experts as though they were priests, the weakness of such feeble arguments would be all the more evident. As it is, creationists are seriously proposing that school children in their early teens be "presented with both sides of the story" and "allowed to decide for themselves", so undercutting the authority of designers of the school curriculum who regard it as their responsibility to decide what children should be taught.

There is a very good reason why children can't themselves decide these issues. Most of what is taught in school is learnt on authority. Children are not presented with the evidence justifying what they learn about the Vikings, geography, foreign cultures and religions; most of the chemistry and physics they learn is not demonstrated before their eyes, and mathematics is not presented with rigorous proofs from first principles, but packaged in persuasive formulations. Most children have neither the time nor ability to make any alternative approach to these subjects feasible, and evolutionary theory is no exception. Now the creationist's arguments typically rest on definite misunderstandings of what the evidence is and what science says about it, and to appreciate the point requires considerable study. To illustrate, one of the creationist's recurring arguments is that evolution conflicts with one of the basic laws of physics, the second law of thermodynamics. Thermodynamics is normally regarded as too abstract a subject to take up in the school curriculum, and most students who encounter it do so first at university. This hasn't prevented creationists brashly claiming that the second law of thermodynamics states that entropy always increases, whereas in biological systems such as an organism, a genealogy including an initial pair of organisms and their descendants, and so on, all display a decrease in entropy over time. Entropy is a difficult concept which can be said to be introduced by a proper statement of the second law of thermodynamics. It is defined mathematically and given various interpretations in different contexts. Suffice it to say here that anyone taking a position on the creationist's claim of inconsistency must be able to assess whether it is based on a correct understanding of the law. Needless to say, it is not. The entropy is defined for a system at equilibrium, and what the second law implies is that a system at equilibrium attains a maximum value of its entropy *for a given energy*. Biological systems do not satisfy this constraint, taking in energy from the sun and food. This doesn't prevent the laws of thermodynamics being applied to biological systems. The basic principles of catalysis governing biochemical pathways, for example, conform to these laws; here the operative constraints are the temperature and pressure rather than constant energy. This is typical of the creationist's claim to parade as members of the scientific community with an alternative view at the centre of internal scientific debate. Disentangling the claims from the rhetoric and treating them as they deserve for an audience of young teenagers is not feasible, even if we were prepared to expose our children to such dishonest propaganda. By the same token, however, we would hope that science is taught in such an undogmatic way that the student eventually comes to appreciate the status of the knowledge acquired during schooling.

The educator is faced with a dilemma. Alternatives to science such as creationism and astrology are easy options—their common failing is that they offer no systematic, disciplined explanations comparable to the substantial and difficult theories they question. They shed no light on questions that orthodox science is unable to answer, and suggest no startling predictions or new fruitful avenues of inquiry. They are not integrated with the rest of science, but stand out as islands desperately clinging onto beliefs grounded in a tradition with no independent support. To illustrate, the English naturalist and member of the Plymouth Brethren, Philip Gosse (1810–1888) was aware that the fossil record studied by Lyell and others seemed to show the earth to be hundreds of thousands, perhaps millions, of years old. To eliminate conflict with his belief that God created the world in 4004 BC, he suggested that earth was created in 4004 BC along with "bogus" geological and fossil records that made it look older. Clearly, the same could equally be said about any other creation story, and certainly provides no support for the 4004 BC date. Contrast this with Cavendish's estimate of the age of the earth based on his calculations of the rate of cooling of volcanic material, which he abandoned when it was realised that radioactivity provided another source of heat. The creationists allow no conflicting evidence to speak against their view. Nor can its supporters complain that the poor state of development of their subject is due to lack of resources. Many talented scientists in the nineteenth century and earlier have espoused their doctrines and attempted to integrate them into a general scientific view. But nothing has come of their efforts. Yet because of the simplistic character of these views, there is no intellectual obstacle of the kind that requires substantial disciplines to be presented systematically with a steady progression of depth and prevents them being presented in full early in the school curriculum. But if this authoritarian approach is necessary, it should also be appreciated that the unorthodox options ultimately offer no real alternative at all.

4.6   Science and Responsibility

This chapter has been much concerned with the moral rectitude of the scientist. It began with issues internal to the practising of science—scientists should be honest, not stealing the results of the labours of others, whether for material gain or renown, by plagiarism, misuse of supervisory relationships, or otherwise failing to acknowledge sources. Authority in the scientific hierarchy shouldn't be abused by favouring personal relations or interests. Results shouldn't be misrepresented by trimming, cooking or forging, and theories shouldn't be promoted as though they were more certain than the evidence and qualified discussion of the topic actually merits. Nor should the investigator shirk from giving as precise an account of their work as possible by overestimating possible error. Leaving such internal matters aside now, what responsibilities does the scientist have towards the community at large?

We have seen something of this in the responsibility to see to it that our children are brought up with some conception of the world as understood in the light of disciplined study and meeting reasonable intellectual standards. Education is also an issue in motivating the investment of large sums of public money in projects of interest to scientists. Public support for the human genome project was encouraged by the promise that it would mean a giant step in the direction of finding cures for human diseases. The suggestion was that this would be a direct application of the new knowledge and we would be reaping these benefits in the fairly near future. Scientists were not to be heard in the media warning that any such application would still lie decades in the future. They were even less forthcoming about the most direct and immediate applications of the new knowledge being the facilitating of genetic testing. Since the most likely uses of genetic testing are ethically highly dubious, failure to alert the public to the dangers of the human genome project and initiate debate is a serious abdication of the scientist's responsibility (Kitcher 2001, pp. 182–92). In some cultures, agents are not regarded as culpable for their omissions. And scientists have all too often sought to avoid blame by claiming that no action of theirs actually triggered the calamity. But if, because of their unique knowledge, the only people in a position to alert the public fail to do so, then there is little hope for conducting society on the basis of well-informed decisions.

The unique positions of scientists as the only people able to understand and be fully aware of possible consequences of their own and their colleagues' work places a responsibility on their shoulders for considering and investigating such possibilities. Prior to the first atomic bomb test in New Mexico there was a worry that a hitherto untried explosive chain reaction in the atmosphere might lead to a chain reaction in the elements constituting the atmosphere. Fortunately, Hans Bethe considered the likelihood of such an outcome beforehand and determined that it was theoretically impossible. No one else could have foreseen this scenario and determined that the risk was negligible.

Cooking the books is as reprehensible in pursuit of public support as it is in seeking recognition and approval of the results of research, perhaps more so since recognition within the scientific establishment can be effectively withdrawn. Returning to a state of ignorance about the human genome, on the other hand, is not possible. All too often, the damage has been done. Knowledge of how to build machine guns, chemical and biological weapons, nuclear bombs, and whatever the weapons industry has in store for us, cannot be rescinded. Knowledge is certainly not of intrinsic, positive value, irrespective of content. Some truths may be of value to the warmonger and weapons manufacturer, but have decidedly negative consequences for the vast majority of mankind. It is difficult to resist the conclusion that some things would be better not known.

But there is a distinct danger of overreacting, shrinking from the truth like the paranoiac afraid to go out for fear of being knocked down. Religious apologists have sometimes argued that the evidence provided by the life and earth sciences refuting religious doctrines such as Christianity has deprived people of the consolations of religious belief. More generally, some have lamented the undermining of traditional values and ways of life by the subversive truths of science, if not going quite so far as the Luddites in early nineteenth century England who destroyed machines of the industrial revolution in the hope of regaining a rural idyll. Beliefs and values such as these are hardly free from their own down side. What of the wars conducted in the name of religion, the oppression of the Inquisition, the cynical use of the power of the church for individual aggrandisement, all at the price of colossal suffering? What human dignity is there in maintaining the creation myth on the grounds proffered by the fundamentalists? Surely, society best serves its members by encouraging them to develop their values and ways of life on the basis of an insight into the best account of the truth about the natural world rather than a gross deception. There may be many particular facts which it does no good to know; but encouraging belief in falsity is hardly a mark of respect for our fellow men. This is not, of course, to excuse the misuse of knowledge and the direction of research into avenues that are clearly detrimental to human well-being. A rational society should promote the formulation and pursuit of worthwhile goals, and hold those who pursue their own ends at the cost of others responsible.

How should blame be apportioned? The terrifying products of the weapons industry derive in part from very general knowledge gained in contexts where these gruesome applications could not be suspected. The famous Einstein equation $E = mc^2$, indicating the energy yield from the loss of mass in nuclear reactions, came to light with the theory of special relativity in 1905, which didn't seem to concern nuclear structure. It was concerned with the theoretical problem of resolving the conflict between Maxwell's electromagnetic theory and Newtonian theory. This general knowledge has been publicly available world-wide since its discovery; and the concept of critical mass is common knowledge. But few countries have actually developed the bomb. Designing bombs is a highly specialised area in which specific technological problems had to be overcome for the purpose of exploiting nuclear physics. Similarly, general knowledge of chemistry and biology is not sufficient to build weapons; specific technological problems have to be solved. Clearly, such programmes specifically designed to produce weapons would not be possible without the complicity of scientists. But politicians and their advisors are equally culpable for instigating such programmes.

How should society react? On the problem of ensuring appropriate expert advice to guide public decision making, there is the idea of ensuring the existence of disinterested informed opinion by the establishment of academic posts protected from political pressure and private influence. This is an old ideal, requiring appropriate conditions of employment and financing of research without strings attached to private interests if the necessary independence is to be secured. There is also a need to protect whistle-blowers if people privy to secrets which should be public knowledge are not obliged to make unreasonable personal sacrifices in going public. As recent cases reveal, this is as much a problem in public life as in private industry. But even supposing the establishment of our independent expertise, lines of communication to the public via a competent journalism in accessible media, and to political decision makers, must be in place. Moreover, the political decision makers must themselves be competent and willing to grasp the issues as well as being amenable to reasonable and widespread opinions from all walks of life and corners of society. It is natural to think in terms of settling science policy on the basis of an ideal of enlightened democracy. Given the complexity of the issues, it is necessary to guard against a tyranny of the ignorant. This calls for some form of representation which is not susceptible to persuasive campaigns of half-truths and outright deception run by interested parties. It must strike a reasonable balance, not representing the interests of some members of society at the cost of inadequately representing the interests of others. Hopefully, with provision for a reasonable level of comfort, the intellectual stimulus and satisfaction of doing one's duty would be proof against corruption as required by any form of democracy, at least under effective principles of openness.

Part II
Philosophies of Science

Chapter 5
Popper: Proving the Worth of Hypotheses

5.1 Popper's Two Central Problems

Karl Raimund Popper (1902–1994) was born in Vienna to upper middle-class parents—his father was a lawyer and his mother a talented musician—who converted to Lutheranism as part of the cultural assimilation of jews. His father took an interest in philosophy, the classics and social and political issues, and had an enormous personal library. Popper was to describe his upbringing as "decidedly bookish". As a young student he was attracted to Marxism, but became disillusioned by what he regarded as pseudo-scientific Marxist historical materialism. He studied mathematics, physics, psychology and philosophy at the University of Vienna for ten years after the First World War, and received his doctorate in psychology in 1928 with a thesis entitled "On Questions of Method in the Psychology of Thinking". The following year he qualified as a secondary school teacher in mathematics and physics. Fearing the rise of Nazism, he worked on his first two-volume book manuscript, *Die beiden Grundprobleme der Erkenntnistheorie*. But he condensed this into *Logik der Forschung* which he published in 1934 (his English translation, *The Logic of Scientific Discovery*, appeared in 1959) and with that he began his career in philosophy. After study leave in England he left Austria in January 1936 with his wife Josefine Anna Henninger (1906–1985) to take up a lectureship at Canterbury College in Christchurch, New Zealand. There he wrote *The Poverty of Historicism* (1944) and his two-volume work *The Open Society and Its Enemies* (1945). Immediately after the war he was appointed to a readership at the London School of Economics and Political Science, where he later became professor and remained until retirement in 1969, remaining intellectually active for the rest of his life. He received many honours during his academic life, was knighted in 1965 and elected a Fellow of the Royal Society in 1976.

Popper regarded himself as an immediate successor of the Vienna circle by reacting against logical positivism in the 1930s, and during his long philosophical career he didn't deviate essentially from the conceptions underlying his original

© Springer Nature Switzerland AG 2020

P. Needham, *Getting to Know the World Scientifically*, Synthese Library 423,
https://doi.org/10.1007/978-3-030-40216-7_5

objections. The grounds on which he criticises logical positivism are to some extent already implicit in what he saw as the two major problems facing the philosopher of science: How can we account for the extraordinary growth of scientific knowledge? and How is the demarcation line to be drawn between what does and what does not count as science? These are epistemological questions. If the logical positivists of the 1920s and 1930s took a small step in the direction of regarding meaning as the fundamental discipline in philosophy, and away from the view which has been dominant since the seventeenth century that epistemology is basic, Popper has done what he could to bring philosophy back on course.

5.2   The Problem of Induction

What Popper sees as distinctive in his epistemology is the attempt to characterise the scientific endeavour as the pursuit of a general methodological procedure. The logical positivists also looked to science for the solution of epistemological problems. But, says Popper, their view of scientific methodology is founded on the principle of induction, and Hume has shown that inductive arguments cannot be justified.

The term "induction" has been explained in Chap. 1 and used on a number of occasions in later chapters. It refers to the kind of argument which musters evidence in support of a conclusion but without deducing the conclusion from this evidence. No matter how strong this argument may be, it always falls short of a deduction. The logical possibility of the premises being all true but the conclusion false—a possibility which deduction rules out—provides a source of uncertainty that has always been a worry, all the more so for those who held knowledge to be certain. But what was just a worry became a more devastating problem for those drawn by David Hume's argument that induction cannot be justified. This, the "traditional problem of induction", has become one of the classic issues in the philosophy of science. It has been taken particularly seriously by empiricists, and Russell described it as the scandal of philosophy.

Newton enunciated four "rules for the study of natural philosophy" (*Principia*, Bk III; Newton 1687, pp. 794–6) in an attempt to systematise the general principles of argument which "conclude with less force" than deductive arguments, and are used to guide the search for laws of nature. These involved ideas of (1) simplicity—laws should be formulated as simply as possible; (2) similarity—similar causes have similar effects; (3) the analogy of nature—sanctioning the inference from observed qualities attaching to all bodies within our experience to the application of these qualities to any bodies whatsoever; and (4) a proviso on the third rule—the results of such inferences are to be regarded as true until such time as fresh evidence turns up exceptions calling for more accuracy. Newton's rules give expression to the non-demonstrative character and uncertainties of inductive inference.

Hume seized on Newton's notion of the analogy of nature—the idea that nature is uniform—sanctioning the belief that phenomena in some as yet unknown or hitherto uninspected realm are similar to, or of the same kind as, phenomena within the realm of our experience. The distant universe is like the near universe; the micro-world is like the macro-world; the future is like the past. Newton's idea gives expression to a general principle of induction, according to which we can infer that regularities observed in the past will continue to hold in the future, and have held all those times in the past even though this has not actually been observed to be the case. Hume asked How can this principle of the uniformity of nature be justified?

Hume understood Newton to be justifying induction by appeal to his notion of the analogy of nature and saying that inductive argument makes tacit appeal to a premise expressing the uniformity of nature. But how, asks Hume, is this premise justified? It cannot be demonstrated, because it is not contradictory to deny it. A world in which it doesn't hold might well be chaotic, but that is not a logical impossibility. Nor can it be justified by appeal to evidence. It is plainly stronger than any set of observations (it is, in fact, stronger than any particular law), which must be complemented by some additional premise to the effect that nature is uniform if the conclusion is to follow. But that assumes what is to be proven, and takes us in a circle.

Hume went on to formulate the problem as a dilemma without appeal to Newton's principle of the uniformity of nature. How can induction be justified? Not, he says, by deduction. However strong our belief in the proposition that the sun will rise tomorrow, our only grounds for believing this are that the sun is known to have unfailingly risen at regular intervals in the past, and that doesn't imply that it will rise tomorrow. No contradiction is involved in asserting that the sun has unfailingly risen at regular intervals in the past, and it will not rise tomorrow.

Failing a deductive justificatory argument, we might appeal to the fact that on the day before yesterday we successfully inferred that the sun would rise yesterday (perhaps not explicitly, but our actions were governed by that belief, among many others). Again, on the day before the day before yesterday, we successfully inferred that the sun would rise on the day before yesterday. And so on. This is tantamount to appealing to the fact that induction has worked adequately in the past, and on that basis inferring that it will continue to work in the future. But this, Hume objects, is circular. It is an inductive inference, projecting a past regularity onto the future. But inductive inference is precisely what is at issue.

The argument is essentially a dilemma. Justification is provided by sound argument, and there are two kinds of argument: demonstrative and non-demonstrative (inductive) argument. This gives us the two horns of the dilemma. But the first horn presents an inadequate justification, and the second horn takes us in a circle. Therefore induction cannot be justified.

Hume concluded that the psychological processes whereby people come to hold beliefs reaching beyond their immediate experience, such as those dealing with the future, cannot be justified by rational argument. The constant conjunction of events of particular kinds establishes a habit which leads the mind to expect an event of one kind when confronted with an event of another. Accordingly, a prediction accords with a past regularity because the regularity conforms with custom and habit. Hypotheses at variance with established habits of mind are rejected. He rejected an occult and wholly mysterious principle of the uniformity of nature, more general than any universal law it could be called upon to justify, and he put his faith in human nature where matters of custom and habit play a central role. This is the scandal Russell spoke of. Surely science is a rational enterprise, and scientific reasoning provides the best kind of rational argument we could hope to find. There must be something wrong with an argument which is claimed to undermine the rational basis of science. But Russell, writing some two centuries after Hume, couldn't put his finger on what was wrong, and many other empiricists have resigned themselves to being able to do no better either.

Popper too held that there is no way of faulting the argument constituting the traditional problem of induction and the task of justifying induction is a hopeless one. Inductive methods should be abandoned and replaced by a more secure methodological principle. Falsification—roughly, the doctrine that we value those theories which have survived serious attempts to prove them wrong—is to fulfil this role, based as it is on the deductively valid principle of modus tollens (from "A implies B" and "not-B" infer "not-A"). He then answers his first question of how we can account for the extraordinary growth of scientific knowledge by talking of science as an "unending quest" involving the replacement of falsified theories with others, which themselves eventually succumb to the same fate, there being no evident reason to suppose that this process will some day come to an end.

5.3   Demarcation

How can a dividing line be drawn between what is to count as science and other, non-scientific activities? Examples of what in Popper's view belong to this latter category include astrology, Marxism, Freudian psychology, Darwinian evolutionary theory and traditional metaphysics. (In later years he modified his harsh assessment of the hypothesis of natural selection; see Popper 1983, pp. 239–46.) The positivists broached this question of demarcation, but took it to be an issue concerning interpretation or meaningfulness, whereas Popper saw it as an epistemological one concerned with grounds for justification. What is characteristic of science is the method by which statements are justified. Once again, falsification plays a central role, and Popper's criterion of demarcation is, broadly speaking, that the attitude towards scientific claims is justified in accordance with methodological principles or rules based on the notion of falsifiability. Popper enunciates a number of such rules which are based on methodological decisions about how to go about accepting and rejecting hypotheses. An example of such a rule is the following.

> Once a hypothesis has been proposed and tested, and has proved its mettle, it may not be allowed to drop out without 'good reason'. A 'good reason' may be, for instance: replacement of the hypothesis by another which is better testable; or the falsification of one of the consequences of the hypothesis. (Popper 1968, pp. 53–4)

The positivists distinguished between two categories of meaningful statements: those true or false in virtue of their logical form (tautologies and contradictions) or reducible to such by definition, and logically contingent statements open to empirical verification. A. J. Ayer presented the creed thus:

> ... it will be shown that all propositions which have factual content are empirical hypotheses; ... every empirical hypothesis must be relevant to some actual, or possible, experience, so that a statement which is not relevant to any experience is not an empirical hypothesis, and accordingly has no factual content ...
>
> It should be mentioned here that the fact that the utterances of the metaphysician are nonsensical does not follow simply from the fact that they are devoid of factual content. It follows from that fact, together with the fact that they are not a priori propositions ... A priori propositions, which have always been attractive to philosophers on account of their certainty, owe this certainty to the fact that they are tautologies. We may accordingly define a metaphysical sentence as a sentence which purports to express a genuine proposition, but does, in fact, express neither a tautology nor an empirical hypothesis. (Ayer 1946, p. 41)

Popper didn't want to go quite so far as to reject metaphysics as meaningless. Metaphysical themes lie behind the thinking of many famous scientists and have played a crucial role in arriving at the formulation of new hypotheses and theories; they cannot be dismissed as nonsensical. Nevertheless, they are not relevant to the real import and significance of a scientific theory, and therefore play no role in the justification of theories. Popper wanted to characterise a scientific theory as one which is open to justification by empirical means, which for him means falsification. And metaphysical propositions, while not to be despised as utter nonsense, are neither tautological nor open to empirical test.

The death blow to the positivists' verification theory of meaning came when it was applied to itself: the verification principle (that all non-tautological significant statements are verifiable) is not itself a tautology. But nor is it verifiable. Clearly the principle is dogmatic metaphysical claptrap by the verificationist's own criterion. Does the application of the demarcation criterion to the methodological principles themselves similarly undermine Popper's philosophy of science? Treated as ordinary statements, they are neither analytic not falsifiable, but at best metaphysics and so irrelevant for justification, which would hardly be an improvement over the positivist's predicament. But Popper doesn't regard methodological principles as ordinary statements with a truth value. They are normative strictures which he says are adopted by decision—conventions whose ultimate justification is pragmatic:

> My only reason for proposing my criterion of demarcation is that it is fruitful: that a great many points can be clarified and explained with its help .... It is only from the consequences of my definition of empirical science, and from the methodological decisions which depend upon this definition, that the scientist will be able to see how far it conforms to his intuitive idea of the goal of his endeavours (Popper 1968, p. 55)

What, ideally, distinguishes the method of science is its self-critical attitude towards its theories. Hypotheses or theories are subjected to searching test by comparing the consequences of the theory with observation and experiment, and the value of a theory to science is measured by its having survived such critical examination. How scientists come by their ideas is another matter—connected with metaphysical views and psychological dispositions of individual scientists which do not form part of the rational enterprise of justification. Rationality comes into the picture first when consequences are deduced from the hypothesis. These consequences are then tested, and if proved false, the original hypothesis is falsified. If, on the other hand, it survives the test, the hypothesis is said to have been *corroborated*.

The great virtue of this view, according to Popper, is that the form of argument involved is deductively valid. If a hypothesis H implies S, but it turns out that not-S, then we can infer with complete confidence that H is false. This is the inference form *modus tollens*. It underlies the notion of corroboration, which is therefore superior to that of verification:

> Nothing resembling inductive logic appears in the procedure here outlined. I never assume that we can argue from the truth of singular statements to the truth of theories. I never assume that by force of 'verified' conclusions, theories can be established as 'true', or even as merely 'probable' (Popper 1968, p. 33).

Thus, it is Popper's contention that there is a fundamental asymmetry between corroboration and verification. Verification, based as it is on induction, has no rational foundation, whereas corroboration, depending as it does on the notion of falsification, has a firm rational foundation in the principle of *modus tollens*.

In practical terms, Popper sees the scientist as starting from hypotheses rather than disparate "observations". Theories give the lead, suggest what is to be investigated, where to look, what sort of thing to look for, how to measure it, and so on. The scientist attempts to corroborate his theories; but it is the theories themselves which direct how they might best be falsified. Popper is sceptical of the traditional empiricist idea that observations provide any sort of starting point, a view which clearly emerges in his criticism of the regularity account of natural laws. According to this view, laws merely express a generalisation, that all so-and-sos are such-and-such, which doesn't distinguish them in principle from any true generalisation, no matter how accidental it may appear to be. If all the coins in my pocket at noon today just happen to be copper, what would then be the true generalisation, "All the coins in my pocket at noon today are copper" is no less lawlike than "All bodies move with the same rectilinear velocity unless acted on by some external force".

Popper begins by pointing out that the observation or experiencing of a regularity between events of kind P and events of kind Q presupposes a prior notion of similarity in terms of which events can be recognised as being of the same kind P or the kind Q. But similarity, he maintains, is a matter of respects from points of view. Given two objects, it is usually not too difficult to think of some respect in which they are similar. Some appropriate universal must, he argues, be specified first before a regularity between two kinds of events can be discussed, and this fact "destroys both the doctrines of the logical and of the temporal primacy of repetitions", which shows "how naive it is to look upon repetition as something ultimate, or given" (Popper 1968, p. 422). Generally, given a finite set of objects or events, there are many hypotheses covering regularities among them. To take a numerical analogy, given a finite sequence of numbers, a rule (a function) can always be found for generating, in addition to the numbers initially given, any number whatsoever as the next number of the sequence.

The message Popper sees in this is that, questions of justification aside, simple enumerative induction cannot be a way of acquiring knowledge of theories. No theory is determined by a finite list of observations. Furthermore, actual physical laws are radically different from simple regularities like "All swans are white". "The problem of showing that one single physical body—say a piece of iron—is composed of atoms or 'corpuscles' is at least as difficult as that of showing that all swans are white". Or again, "we cannot show, directly, even of *one* physical body that, in the absence of forces, it moves along a straight line; or that it attracts, and is attracted by, one other physical body in accordance with the inverse square law" (Popper 1968, p. 422). The laws at issue here "transcend all possible experience". Popper says that "inductivists" acknowledge the difficulty of accounting for such "structural theories" and solve the problem by sharply distinguishing observational generalisations and abstract theories. Theoretical statements are then treated as instruments, making no genuine assertions about the world. Popper sees no good grounds for drawing such a rigid distinction. "We are theorising all the time" is his slogan. Abstract theories are appealed to in the most ordinary, everyday terms which are first devised for the purpose of formulating hypotheses and theories. Theories are implicated in the most basic vocabulary; even saying something is a swan involves the attribution of "properties which go far beyond mere observation—almost as far as when we assert that it is composed of corpuscles" (Popper 1968, p. 423).

This is how Popper sets up the debate with the positivists and contrasts his views with theirs. He understands an inductive argument to be one which "passes from *singular statements* ..., such as accounts of the results of observations or experiments, to *universal statements*, such as hypotheses or theories" (Popper 1968, p. 27). The term "verification" has a correspondingly narrow sense in his usage, referring to justification by enumerative induction. It is the validity of arguments of this kind that Popper thinks are called into question by what he takes to be Hume's incontestable argument, and which must be replaced in an adequate methodology of science. His central notion of falsifiability is not quite so simple as the above references to *modus tollens* might suggest, however, and Popper allows himself to incorporate several nuances.

5.4 Falsifiability

Popper's notion of *falsifiability* involves the satisfaction of certain methodological requirements, taken up presently, and certain logical requirements which are discussed first. An empirical theory should be logically falsifiable; i.e. it must be at least logically possible to test it. The outcome of a test—an experiment or an observation—is a categorical singular statement. But a general statement of the form *All Ps are Qs* implies at best only a conditional singular statement *If a is P, it is Q*, where "a" is a name denoting some specific individual. In order to derive a categorical singular statement, *a is Q*, from a general statement, there must be some further singular statements—sometimes called initial conditions—in conjunction with which the general hypothesis does imply a singular statement. Thus, *All Ps are Qs* and *a is P*, for example, imply *a is Q*. More specifically, a *falsifiable* hypothesis must imply a singular statement distinct from every initial condition.

A hypothesis is thus falsifiable with respect to some given initial condition. Popper recognises this (1968, pp. 75–6) when he says that the initial conditions are themselves also empirical hypotheses in the sense that they too are equally subject to falsification. He makes use of this observation in dealing with metaphysical postulates such as "All events have a cause" and "Matter cannot be created from nothing". In order to exclude such statements, which would count as falsifiable hypotheses unless the definition is modified, he restricts the singular statements which can be said to falsify a hypothesis to what he calls *basic statements*. These are singular statements which can actually serve in practice as premises in a *modus tollens* inference to falsify a hypothesis. His idea is that an empirical theory should enable us to deduce more empirical singular statements than follow from the initial conditions alone. From "All events have a cause" and "A catastrophe is occurring here", for example, we can deduce "This catastrophe has a cause" (p. 85). But this statement contains no more empirically significant information than is contained in the initial condition, and certainly no singular statement reporting experimental results implies its negation. Popper goes on to define a falsifiable theory as one which divides basic statements into two non-empty subsets according to whether they are consistent or inconsistent with the theory. The former are permitted by the hypothesis, whilst the latter are potential falsifiers. What a theory asserts, then, on Popper's view is that there are potential falsifiers, but that they are all false. The empirical content of a hypothesis is defined as the set of its potential falsifiers. The theory says nothing about the basic statements it allows, neither that they are true nor that they are false. Falsifiability means here simply that there are some statements which the hypothesis doesn't contradict.

Finally, a theory is not usually regarded as falsified by just one conflicting observation. There must be some reproducible effect which refutes the theory. We therefore have to talk about a falsifying general hypothesis, itself subject to the requirement of being empirically testable by recourse to accepted basic statements, although it would normally be a comparatively low-level empirical hypothesis in relation to the theory at issue (Popper 1968, p. 86).

The characterisation of falsifiability goes beyond purely logical criteria. Methodological rules come into the picture, and it is high time to observe that

> First a supreme rule is laid down which serves as a kind of norm for deciding on the remaining rules, and which is thus a rule of higher type. It is the rule which says that the other rules of scientific procedure must be designed in such a way that they do not protect any statement in science against falsification (Popper 1968, p. 54).

This rule is directed against the doctrine of conventionalism as developed by Henri Poincaré (1854–1912) at the turn of the twentieth century, according to which deep-rooted laws like Newton's laws of mechanics are not synthetic (factual) statements. They bear closer resemblance to definitions than to empirical statements, and any prima facie counter evidence can always be construed so as not to conflict with them. Poincaré considered them to be conventions which we adopt because of the simplicity they impose on our experience. Popper concedes that "it is impossible to decide, by analysing logical form, whether a system of statements is a conventional system of irrefutable implicit definitions, or whether it is a system which is empirical in my sense" (Popper 1968, p. 82). His criterion of demarcation thus leads him to view an empirical theory not just as a set of sentences. "Only with reference to the methods applied to a theoretical system is it at all possible to ask whether we are dealing with a conventionalist or an empirical theory" (Popper 1968, p. 82). We make a decision not to save our system by any kind of conventionalist strategy such as resorting to ad hoc hypotheses, changing the coordinating definitions specifying the empirical content of a theory, or casting doubt on the competence of the experimenter or theoretician (1968, p. 81). It must be clear in advance what would be accepted as grounds for falsification.

5.5 Ad Hoc Hypotheses and Scientific Progress

A general statement is only falsifiable with respect to some given initial conditions. The rule just mentioned means that such auxillary hypotheses required for the deduction of testable consequences of a theory should not diminish the *degree of falsifiability* or *testability* of the system. This notion of degree of testability does heavy duty in Popper's account of empirical science and lies at the heart of his methodological rules. We are to select bold hypotheses making substantial claims about the world rather than timid theories which run very little risk of being falsified because they say so little. The more a theory claims, the greater the chances of its being falsified. Yet if such a bold theory survives our best efforts to falsify it, it is of far greater value to us than a more circumspect hypothesis which easily survives testing. Bolder theories are characterised as those which have a greater degree of falsifiability. Auxillary hypotheses should not be chosen in such a way that the degree of falsifiability of the system as a whole is reduced. This is thus a methodological rule which excludes ad hoc hypotheses, an ad hoc hypothesis being precisely one which makes the degree of falsifiability zero or very low.

Unfortunately, this directive is of little value in identifying a hypothesis as ad hoc. Consider the following example. In 1610 Galileo published his *Starry Messenger* where he reports certain observations of the moon made with his newly invented telescope which refuted, so he claimed, a received view of the time. According to Aristotle, all heavenly bodies (including the moon, but not, of course, the earth) are perfectly spherical with a smooth surface. What Galileo claimed to have done was to observe mountains and craters on the moon's surface, and even to have estimated the height of one of the peaks to be four miles, all of which suggested the moon's surface was just as irregular as the earth's. These observations threatened the Aristotelian view of the division of the universe into a super- and a sublunar region, and some of Galileo's contemporaries leapt to its defence. Their counter claim was that any apparent irregularities were encased in transparent crystal whose surface—the moon's real surface—is smooth. Galileo rose to the challenge. In a letter complying with a request for his comments on the objection emanating from a Cardinal Joyeuse, Galileo sought to disarm all hypotheses formulated with the sole aim of supporting Aristotle and Ptolemy. He granted, for the sake of the argument, the postulation of a transparent material surrounding the moon, but added that "this crystal has on its surface a great number of enormous mountains, thirty times as high as terrestrial ones, which ... cannot be seen by us". Alternatively as Drake (1978, pp. 168–9) puts it, the earth might, by parity of argument, be defined to include its atmosphere and claimed to be perfectly spherical, again eliminating the distinction between the surface of the earth and the moon, but this time by applying what was said about the moon to the earth.

Galileo's play on the invisibility of the theory-saving crystalline substance seems to indicate quite clearly that its postulation is merely an ad hoc device. But clear as this may seem, it is not easy to generalise into a uniformly applicable strategy. Compare the following, more recent example.

It has been known since the end of the nineteenth century that certain atomic nuclei emit beta-radiation, which comprises a beam of electrons (negatively charged particles). In the first decades of the twentieth century, it was thought that beta-decay, in which atomic nuclei disintegrate with the emission of beta-rays, occurs when a neutron in the atomic nucleus changes into a proton (a positively charged particle) and an electron. The electron is emitted and the parent nucleus gains a proton, which means that it becomes a different chemical element. The kinetic energy of the electron comes from what is left of the difference in rest masses of the parent and daughter nuclei over and above its own rest mass in accordance with the Einstein equation $E = mc^2$, and is a quite definite quantity. Now, in virtue of the fundamental physical principle of the conservation of energy according to which energy (including mass) cannot be created or destroyed, all electrons should emerge with the same very definite kinetic energy defined by the rest masses of the nuclei and the electron. Moreover, according to another fundamental principle—the law of the conservation of linear momentum—the daughter nucleus should recoil in the direction opposite to that in which the electron emerges. But these test implications are violated by the experimental results, and the two conservation laws accordingly falsified. (A third law of the conservation of angular momentum is also falsified.)

Unwilling to abandon these classical principles, Pauli suggested in 1933 that the kinetic energy missing from the emerging electrons could be accounted for by postulating that a new kind of particle is emitted along with the electron and which Fermi, in the course of developing the suggestion, dubbed the neutrino. These particles could not have any charge, for otherwise the law of conservation of electric charge would be violated. But this meant that none of the methods for detecting particles which rely on particle charge could be used to verify the existence of neutrinos. Neutrinos must also be assigned zero rest mass, and therefore also the velocity of light, which would suggest that they resemble photons such as gamma rays. But none of the easily-observable effects which can be used for detecting photons work with neutrinos. They are practically invisible! The situation prompted the revival of the age-old doctrine of instrumentalism, lately advocated by the positivists, according to which theoretical entities don't really exist and theoretical claims are not true or false but simply more or less useful in efficiently organising observational data. "[I]t doesn't make any sense," as one physicist put it, "to argue about questions like the reality of the neutrino, or for that matter of the electron or proton. I would hold such discussions to be meaningless". The job of the theorist is "to find some trick way of correlating the data in as big lumps as possible" (Dancoff 1952, pp. 139, 140).

In fact, neutrinos were finally detected in 1956. Their properties are such that they interact very rarely with matter. The earth, for example, provides virtually no obstacle to solar neutrinos. As Griffiths (2004, pp. 24–6) puts it, they "interact extraordinarily weakly with matter; a neutrino of moderate energy could easily penetrate a thousand light-years(!) of lead". The "invisible" particles were finally detected, some thirty years after they were first postulated. But in the 1930s, when there seemed to be no prospect of detecting neutrinos, the situation was very much like that at the beginning of the seventeenth century when Galileo dismissed the arbitrary postulation of an invisible substance in defence of Aristotelian principles. Compare this with a third example, the explanation of a discrepancy between Uranus' orbit and that predicted by Newtonian mechanics which exercised the minds of Leverrier in France and Adams in England during the last century. They chose to account for the anomaly by postulating the existence of a planet whose gravitational influence would account for the observed orbit of Uranus. As it turned out, when they pointed their telescopes in the direction of the night sky where their calculations suggested the planet ought to be, they discovered Neptune (or rediscovered it after Galileo if Drake and Kowal (1980) are to be believed). But if they hadn't, how far would they have gone in postulating untoward atmospheric effects, intervening stellar gas clouds, and so forth (cf. Lakatos 1978, p. 17)? A planet similarly postulated to account for anomalies in Mercury's orbit was actually named "Vulcan" before being sighted. It still hasn't been sighted yet.

Popper's rule, that theoretical systems should not be protected from falsification, would seem to work equally well against the neutrino hypothesis as against the invisible lunar material. According to another of Popper's methodological rules, a theory should not be abandoned until a better theory is available. This would explain the rationale behind Pauli's postulate as that of hanging on to the classical principles for a sufficiently long time to allow the detection of the neutrinos. But equally in Galileo's case, it can be argued that the theory which was to replace the Aristotelian view was not developed until sometime later in the seventeenth century (Newton's synthesis). And so again it seems no clear-cut methodological distinction has been drawn.

Interestingly enough, Popper cites the Lorentz-FitzGerald contraction hypothesis (to the effect that contraction in length occurs in the direction of movement through the ether just sufficient to account for the inability of the famous Michelson-Morley experiment of 1887 to detect movement relative to the ether) as an example of an ad hoc hypothesis. The contraction hypothesis was put forward before the special theory of relativity appeared, when it was apparent that Maxwell's laws of electromagnetic radiation, the traditional Galilean kinematics, and the traditional restricted principle of relativity or homogeneity according to which the laws of physics are indifferent to the constant velocity of the system from whose vantage point they are applied, are jointly inconsistent. Something must be given up. But it is not difficult to understand resistance to abandoning the Galilean principle of addition of velocities according to which if B observes C moving at speed v_1 to the right and A sees B moving at v_2 to the right, then A would observe C moving at $v_1 + v_2$ to the right. Nevertheless, Maxwell's laws imply that light moves with a definite velocity c in empty space. But the Michelson-Morley experiment showed that the velocity of light as measured on the moving earth does not vary as Galilean kinematics requires. The distinction between absolute and relative motion has, it seems, no experimental significance and light has the same speed relative to A and to B. So if the indifference of laws to frame of reference is to be maintained, the Galilean addition principle is contradicted. But is it so clear that the contraction hypothesis is more arbitrary or ad hoc than abandoning Galilean addition of velocities? Poincaré did come to this view, accepting the modification of the time axis which the Galilean transformations leave unchanged. But he resisted this view at first, and it doesn't require too much imagination of the modern reader to understand that it was a surprise to learn that the Galilean additivity principle doesn't go hand in hand with relativity. The naive view remains, in fact, the Galilean one, that the time axis is unaffected by change of viewpoint. From the perspective of special relativity it is easy to debunk the incorrect hypothesis as ad hoc; but this smacks of a judgement *post factum* and not one based on timeless methodological criteria. (Bohm (1996) explains that the Lorentz contraction arose in a natural way from his theory of electrons, but subsequent examination of Lorentz's theory showed that it rendered systematically impossible observation of the motion relative to absolute space that it built upon. See Hunt (1991, pp. 185–97) for a more detailed account of FitzGerald's reasoning.)

Consider another example. Avogadro's hypothesis that equal volumes of all gases at the same temperature and pressure contain the same number of particles, presented in 1811, was essential to Cannizzaro's resolution of the problem of determining chemical formulas in 1858. But the evidence seemed at first to count decisively against the hypothesis. In the first place, densities were thought to raise a problem. For example, although oxygen is denser than steam, a particle of steam must, it seemed, be heavier than one of oxygen since it contains more than just oxygen. So it was reasoned that one unit volume of steam would contain a smaller number of steam particles than one unit volume of oxygen would contain of oxygen particles. Furthermore, the hypothesis seemed to be at odds with reacting volumes. Just two volumes of steam result when two of hydrogen combine with one of oxygen, for example. It was assumed that gaseous elements are simple, leading Berzelius to express the reaction in the form

$$2H + O \rightarrow H_2O.$$

This implies, if anything, that a volume of steam comprises only half as many particles as the same volume of hydrogen. Avogadro's interpretation was different, in line with his own hypothesis. He proposed that hydrogen and oxygen comprise diatomic molecules, leading him to represent the reaction in modified Berzelius notation as follows:

$$2H_2 + O_2 \rightarrow 2H_2O.$$

Since this is how the reaction is formulated today it may strike us as the better proposal. But Avogadro had no way of independently confirming the diatomic structure of hydrogen and oxygen, and we should be wary of allowing our contemporary knowledge to lead us to view Avogadro's proposal as anything but an ad hoc manoeuvre to save his hypothesis. As Frické (1976) points out, were two volumes of hydrogen to combine with one of oxygen yielding n volumes of steam, then Avogadro would have represented the reaction by

$$2H_n + O_n \rightarrow nH_2O,$$

whatever the value of n. For his part, Berzelius advocated an electrical theory of chemical combination according to which quantities of different elements are held together in a compound by a balancing of the negative charge of one kind of matter with the positive charge of another. Consequently, like atoms would repel one another and diatomic molecules would be an impossibility (as they would, for a different reason, on Dalton's theory; see Sect. 7.3.2). Berzelius therefore had good reason to reject Avogadro's hypothesis (Nash 1957; Frické 1976; Needham 2018, pp. 351–4).

Should Cannizzaro have followed Popper's advice? Should the fact that Avogadro's hypothesis had already been discredited as an ad hoc defence (and subsequently falsified) have sufficed to permanently ban it from the domain of serious scientific consideration? It is beginning to look as though Popper's attempt to provide criteria of demarcation is at odds with his attempt to provide an account of the extraordinary growth of scientific knowledge.

Popper, like the positivists, is not dealing with what they called the context of discovery. He is interested in "the logic of knowledge ... [which] consists solely in investigating the methods employed in those systematic tests to which every new idea must be subjected if it is to be seriously entertained" (Popper 1968, p. 31). This he distinguishes from the investigation of the psychological processes involved in stimulating the inception of an original hypothesis. "There is no such thing as the logic of having new ideas" (Popper 1968, p. 32). Questions of justification are his only concern. He is proposing a methodology which a scientist should use in justifying theories. The physicist Herman Bondi, for example, used to argue for the steady state theory by appeal to Popper's methodology. According to the steady state theory, atoms of hydrogen are spontaneously created in space at such a rate as to maintain the average density of the universe as it continues to expand. The creation of matter from nothing conflicts with deep-seated intuitions and long-standing principles of metaphysics, not to mention classical physics. But Bondi, following Popper, regards these as psychological considerations which have no bearing on the question of whether we are justified in accepting the theory. The theory has a high degree of falsifiability and has been fruitful in developing theories of the creation of the elements and the development of stars. As it turned out, the consensus of scientific opinion has abandoned the theory in view of the observation of background radiation which could only have come from the big bang. (It is not clear that the steady state theory was abandoned because falsified, however. It just didn't imply the presence of background radiation, whereas the big bang theory does and was therefore verified by the observation.)

Clearly, if Popper's methodology is to stand as a set of norms aloof from the beliefs scientists are inclined to hold, a great deal of its point would be lost if we can only judge in retrospect whether his rules are followed in a particular case. They are to establish contact between the theory and empirical reality where otherwise Popper sees only metaphysics and psychology. But his directives give us no way of singling out ad hoc hypotheses beforehand, and therefore give no clue about what actually directs the progress of science. Popper's student Imre Lakatos was very much aware of difficulties such as these, but girded by the important social and political implications of distinguishing between and supporting science and pseudo-science (Lakatos 1978, pp. 1–7), he persisted with the attempt of his mentor to develop a demarcation criterion. Unfortunately, his conception of scientific research programmes fares no better in providing an explicit criterion. But it reflects the illusion of thinking that scientific methodology can be captured by a simplistic slogan.

5.6   Degree of Falsifiability or Induction by Another Name?

The progress of science, on Popper's view, is marked by a series of bold, imaginative hypotheses which challenge earlier views and offer the hope of explaining a wide range of phenomena. This is not to be construed as meaning that these hypotheses say a great deal about what is true in nature, but rather that they have a large number of test implications and therefore run a correspondingly large risk of being falsified. Popper contrasts bold hypotheses with timid modifications which do not readily open themselves up to the possibility of being falsified and which he maintains are of little value in science. What we are interested in is, rather, a high degree of corroboration, which means not simply the mere survival of attempts at falsification, but the survival after the most ingenious and wide-ranging attempts at proving the hypothesis wrong. Degree of corroboration is a mark of susceptibility to falsification, and this in turn is measured by degree of falsifiability.

Popper contrasts his notion of degree of corroboration with that of the degree of verification, which he sees as a completely misguided concept with little or no bearing on actual scientific practice. He takes J. M. Keynes, in his *A Treatise on Probability* (1921), as a major and typical proponent of verificationism, and quotes disapprovingly Keynes' view that the value of prediction in science is overrated. What, on Keynes' view, is important for the acceptability of a hypothesis is the number of instances examined and the analogy between them; "whether a particular hypothesis happens to be propounded before or after their examination is quite irrelevant" (quoted by Popper 1968, p. 272). This view leads Popper to wonder "why we should ever have to generalise at all". The programme of inductive logic suggests that there is no reason at all for constructing theories and hypotheses. Significance is assigned to hypotheses only to the extent that they can be justified by experience, which means "that the content of the theory must go as little as possible beyond what is empirically established" (loc. cit.). But in that case we might as well rest content with the bare statement of the evidence.

In order to clarify his own position, Popper compares the two hypotheses "All crows are black" and "The charge on the electron is that determined by Millikan".

> Although in the case of a hypothesis of the former kind, we have presumably encountered many more corroborative basic statements, we shall nevertheless judge Millikan's hypothesis to be the better corroborated of the two. (Popper 1968, p. 267)

Degree of corroboration is thus not simply determined by the number of corroborating instances; Popper's proposal is that it depends rather on the severity of the various tests to which the hypothesis can be subjected, which in turn depends on the degree of testability or falsifiability of the hypothesis. Popper expresses this idea somewhat paradoxically by saying that the degree of corroboration—a measure of testability—is inversely proportional to probability. The sense of probability Popper has in mind is a special one he calls logical probability and which he identifies with what Keynes called a priori probability. It is something a statement has "by virtue of its logical form" (Popper 1968, pp. 118–9). Intuitively, a statement is highly probable in this sense if its chances of being true are great. This means that the statement is a cautious one involving, by virtue of its restricted content, as little unnecessary risk as possible of turning out to be false. Timid and restrictive theories hardly extending beyond the observations on which they are based, or unspecific statements so vague or unspecific as to be almost certainly true, are highly probable.

In order to talk of degree of falsifiability some method of comparing theories is needed. One method of comparison which Popper considers is to look at the way sets of potential falsifiers—basic statements which if true would falsify the hypothesis—fall into subsets of one another. When one set of potential falsifiers is a subset of the other, the corresponding hypotheses can be compared with respect to degree of falsifiability. In this case, H_1 is said to be more falsifiable than H_2 if and only if the set of potential falsifiers for H_2 is a proper subset of the set of potential falsifiers for H_1. Where these sets are identical, the corresponding hypotheses are equally falsifiable. Intuitively, a theory is more falsifiable the more potential opportunities there are to refute it. Elaborating on this definition, Popper points out that if H_1 entails H_2, then intuitively H_1 says at least as much as does H_2. Correspondingly, it is natural to require where H_1 entails, without being equivalent to, H_2, that the empirical content (defined by Popper as the set of potential falsifiers) of H_1 is greater than that of H_2 and so H_1 is more falsifiable than H_2. Logically equivalent hypotheses are equally falsifiable. Some further constraints are needed to deal with the possibility of metaphysical statements occurring as part of these hypotheses which logically entail statements which might not be open to testing. But the basic idea is clear, and we can see that more universal hypotheses (those with less restrictive antecedents) and more precise hypotheses (those with more specific consequents) are accordingly more falsifiable compared with the less universal or less precise hypotheses they imply.

Once some such notion of degree of falsifiability is to hand Popper is in a position to make another methodological decision, "sometimes metaphysically interpreted as the principle of causality—to leave nothing unexplained, i.e. always to try to deduce statements from others of higher universality" (Popper 1968, p. 123). Popper thus thinks himself able to provide a rationale for the fact that natural science has pursued the course of striving after ever more general theories which he sees Keynes, and by implication verificationists in general, as being lamentably unable to account for. This actual demand on the part of scientists for the highest degree of generality and precision can be reduced to the demand, or rule, that preference should be given to those theories which can be most severely tested (1968, p. 123).

The weakness of verificationism as Popper defines it—confirmation by simple enumerative induction—is effectively demonstrated by this critique. Nevertheless, in developing what he regards as a more adequate alternative position he has strayed a long way from the humble beginnings which warranted the statement (quoted more fully in Sect. 5.1) that "nothing resembling inductive logic appears in the procedure outlined here". Let us re-examine his initial claims in the light of his more fully-developed views. Consider first the asymmetry claim, that a hypothesis can never be conclusively verified, but falsification is conclusive, based as it is on purely logical reasoning. This is true with reference to hypotheses of the elementary general form so far considered, but not of purely existential hypotheses such as "There are atoms", which no finite list of data can contradict. Nor is it true of hypotheses of only a slightly more complex kind involving both existential and universal quantification, such as "Every substance is soluble in some solvent" and "You can fool some of the people some of the time" (Hempel 1965, p. 40). Bear in mind here that such mixed quantification often lurks beneath the surface of everyday statements; "All men are mortal", for example, is reasonably understood to mean "For every man, there is some time when he dies" (Quine 1974). Falsification of "Every substance is soluble in some solvent" requires finding a substance which is insoluble in *every* solvent, and so is no more conclusive than verification of "All swans are white".

Secondly, the above considerations implicitly reveal an acknowledgement that there is a limit to how far purely logical considerations can take us. Logic can only tell us that a given hypothesis is inconsistent with a given set of sentences reporting the empirical data. In general there will remain a large number of alternative hypotheses each consistent with the data. The problem of scientific inference is to try to find in the data some grounds for preferring one above the rest, calling for further considerations to be taken into account in order to provide guidelines for the selection of hypotheses. This is indeed what Popper intends with his notion of degree of corroboration. He directs us to select bold hypotheses—those with a high degree of corroboration. But in so doing he exceeds the information available in the accepted basic statements and appeals to a non-deductive form of inference. This may be a more complex form of inference than enumerative induction and therefore not, on Popper's use of the term, an inductive inference. It is, nonetheless, a form of non-deductive inference which is inductive in the sense of the dilemma constituting Hume's problem of induction. Hume's argument therefore applies to falsificationist methodology with exactly the same force as it does to enumerative induction; Popper has neither solved nor circumvented the problem of induction.

5.7 Verisimilitude

Popper has in more recent years tempered his scepticism with a glint of optimism. No theory can, he thinks, be justifiably believed and all the theories we use are false. But progress is possible nevertheless. We can't expect to replace a false theory with a true theory, but the new theory can at least be less false. When choosing between false theories we can select that which has the greatest verisimilitude—"the idea of better ... correspondence to truth or of greater ... likeness or similarity to truth" (Popper 1969, p. 233). To illustrate,

> Newton's theory allowed us to predict some deviations from Kepler's laws. Its success in this field established that it did not fail in cases which refuted Kepler's: at least the now known falsity-content of Kepler's theory was not part of Newton's, while it was pretty clear that the truth-content could not have shrunk, since Kepler's theory followed from Newton's as a 'first approximation'.
>   ... Ultimately, the idea of verisimilitude is most important in cases where we know that we have to work with theories which are at best approximations—that is to say, theories of which we actually know that they cannot be true. ... In these cases we can still speak of better or worse approximations to the truth (and we therefore do not need to interpret these cases in an instrumentalist sense) (Popper 1969, p. 235)

Popper sketches two ways of defining verisimilitude, one a quantitative measure of degree of truth, the other a qualitative concept applicable only to theories which are comparable in a sense explained shortly. Taking up the latter first, the problem is to make precise the idea of one theory's having greater truth content than another. Popper's leading idea derives from Tarski's so-called semantic definition of truth from the 1930s. The details, which are quite technical, won't be pursued here. But it is important to understand that Popper takes Tarski's work to show how an objective definition of truth as correspondence to the facts can be given.[1] The notion of verisimilitude, built upon these foundations, is equally an objective concept. It is not an epistemological concept such as the estimation of degree of truth content. I may well not know whether one theory has a higher degree of verisimilitude than another—"I can only guess. But I can examine my guess critically, and if it withstands severe criticism, then this fact may be taken as a good critical reason in favour of it" (Popper 1969, p. 234).

A theory is understood, in the spirit of Tarski, as a set of sentences. Let $Cn(A)$ denote the set of logical consequences of a theory A. (Tarski gives a definition

---

[1] For a discussion of this claim, see Fernández Moreno (2001).

of this according to which any sentence B is in $Cn(A)$ iff each model for A is a model for B; i.e. if, and only if, every way of representing A as true represents B as true.) All theories have infinitely many consequences; nevertheless, it is possible for one such set, $Cn(A)$, to be a proper subset of another, $Cn(B)$. This is written $Cn(A) \subset Cn(B)$, and means that every sentence in $Cn(A)$ is also in $Cn(B)$ but not conversely—there is some sentence in $Cn(B)$ not in $Cn(A)$. Now, two theories are comparable if the one consequence set is included in the other. $Cn(A) \subset Cn(B)$ means that B has a greater content—it says more than—A. But that doesn't suffice for greater verisimilitude. The content of Kepler's theory might be increased by adding to it every sentence of *Alice in Wonderland*, but its truth content wouldn't have increased one iota. Popper distinguishes between true and false consequences—sentences which are both consequences of the theory and true, and sentences which are consequences and false. He then goes on to say that a theory B is closer to the truth than a theory A if it has either a greater truth content without having a greater falsity content, or it has a lesser falsity content without having less truth content. In symbols, if A_T is the set of true consequences of A (A's truth content), and A_F its falsity content, and similarly for B_T and B_F, then Popper defines

> A has less verisimilitude than B if, and only if, either $A_T \subset B_T$ & $B_F \subseteq A_F$, or $B_F \subset A_F$ & $A_T \subseteq B_T$,

where "$A_T \subseteq B_T$" means "either $A_T \subset B_T$ or $A_T = B_T$".

Unfortunately, Miller (1974) and Tichy (1974) showed that if B is false, then A does not have less verisimilitude than B (i.e., only true theories have more verisimilitude than any theory on Popper's definition), which defeats the point of Popper's notion.[2] Elie Zahar, a philosopher close to Popper, takes the view that "Precisely because Popper's definition closely captures the intuitive notion of verisimilitude, Miller's negative result constitutes to my mind a severe setback for fallibilist realism as a whole" (Zahar 1983, p. 167). Worrall (1982, p. 228) was still hoping that "a reasonable characterisation of empirical verisimilitude (of which successive theories may have had ever more) might be rescued from these difficulties. But the idea ... [of] increasing overall verisimilitude seems to me ... generally and intuitively unsound".

On the other hand, despite this unfortunate result Popper's general quest for a notion of verisimilitude has retained its attraction for some philosophers, not least

---

[2]Tichy's proof runs as follows. Suppose B is false. (i) Assume $A_T \subset B_T$. Then for some true sentence τ, $\tau \in B_T$ and $\tau \notin A_T$. To say that B is false means that there is a false sentence $f \in B_F$. Since f is false, so is the conjunction $f \wedge \tau$, in which case $f \wedge \tau \in B_F$. But $f \wedge \tau \notin A_T$; for otherwise $\tau \in A_T$, contradicting what was said about τ. Hence $B_F \nsubseteq A_F$ and A does not have less verisimilitude than B. (ii) Assume $B_F \subset A_F$. Then for some false sentence φ, $\varphi \in A_F$ and $\varphi \notin B_T$. Again, since B is false there is a sentence $f \in Cn(B)$ which is false. Since f is false, the disjunction $\sim f \vee \varphi$ is true. Then $\sim f \vee \varphi \in A_T$. But on the other hand, $\sim f \vee \varphi \notin B_T$; for otherwise $\varphi \in B_T$, since $f \in Cn(B)$, in contradiction with the assumption. Hence $A_T \nsubseteq B_T$ and again A does not have less verisimilitude than B. For both alternatives in Popper's definition, then, a false theory cannot have more verisimilitude than another theory.

Tichy and Miller themselves, who have sought to develop alternative approaches not directly based on the notion of a theory's consequences as in Popper's definition. Perhaps the most thorough treatment is due to Ilkka Niiniluoto. Technicalities preclude the presentation of details here (but see Brink (1989) for an excellent overview). Interest turned to a quantitative account. It is generally acknowledged, however, that there is a certain arbitrariness of choice of measure from several possible measures of the nearness of a hypothesis to the truth, and criteria distinguishing the best measure have not been forthcoming. Moreover, as Miller has pointed out, the measure is so closely tied to a definite linguistic framework that the verisimilitude ordering is not preserved under the most innocuous translations, which is inappropriate if the measure of verisimilitude is to provide a measure of scientific progress. Progress is often achieved by shifting to a better linguistic framework with the introduction of new concepts (and perhaps the dismantling of older ones), and this is therefore beyond the scope of language-dependent measures of verisimilitude. An interesting further approach due to Schurz and Weingartner (1987) returns to the notion of consequence, seeking to capitalise on the idea that scientists may not want all the logical consequences of a theory, as understood in classical logic, but only those which are relevant. For example, if q is a consequence of p, so, according to classical logic, is $q \vee r$, for arbitrary r, although this seems irrelevant ("$\vee$" expresses inclusive disjunction ("or"), so that $q \vee r$ is true if either one or both of q and r are true and false otherwise). Similarly, it seems superfluous, having established p and q, to go on to conclude $p \wedge q$ ("$\wedge$" expresses conjunction ("and"), so that $p \wedge q$ is true if both of p and q are true and false otherwise). Tichy's and Miller's proofs depend on such irrelevant and superfluous consequences, which Schurz and Weingartner filter out and so reinstate at least the spirit of Popper's original definition. It does seem to be true that scientists don't willingly subscribe to the literal truth of all the assumptions and consequences of continuity entailed by the use of mathematical analysis. But Schurz and Weingartner's methods give no real clue about how such unwanted results can be coherently sifted out. A further weakness is that their approach remains that of amassing true sentences whilst ensuring that they are not overwhelmed by false ones. A theory like Newton's combined theory of mechanics and gravitation may be completely false, however, and yet each false claim may be closer to the truth than any claim of a competing theory. Intuitively, being a better shot at the truth doesn't turn on the number of hits (true sentences), but on how good an approximation it is to the truth. As we will see, some such conception would be closer to that underlying Duhem's view of scientific progress.

Independently of the specific formal features of any approach, the appropriateness of the general idea that verisimilitude is a measure of progress has also been questioned. Wachbroit (1986) suggests that the heliocentric theory of Aristarchus—"the first Copernicus"—might be considered nearer the truth than were its contemporary geocentric theories in the third century BC. But it is far from clear that its rejection then should be deemed regressive: upholding it may well not have led to a Kepler, a Galileo and a Newton doing what Kepler, Galileo and Newton actually did some two millennia later.

Chapter 6
Duhem's Continuity Thesis

Scientific progress has often been compared to a mounting tide; applied to the evolution of physical theories, this comparison seems to us very appropriate, and it may be pursued in further detail.

Whoever casts a brief glance at the waves striking a beach does not see the tide mount; he sees a wave rise, run, uncurl itself, and cover a narrow strip of sand, then withdraw by leaving dry the terrain which it had seemed to conquer; a new wave follows, sometimes going a little farther than the preceding one, but also sometimes not even reaching the sea shell made wet by the former wave. But under this superficial to-and-fro motion, another movement is produced, deeper, slower, imperceptible to the casual observer; it is a progressive movement continuing steadily in the same direction and by virtue of it the sea constantly rises. The going and coming of the waves is a faithful image of those attempts at explanation which arise only to be crumbled, which advance only to retreat; underneath there continues the slow and constant progress whose flow steadily conquers new lands, and guarantees to physical doctrines the continuity of a tradition. (Duhem 1954, pp. 38–9)

6.1 Bibliographical Sketch

Pierre Maurice Marie Duhem was born in Paris on 9 June, 1861, into a family with deeply-rooted Catholic beliefs and a strong allegiance to the monarchy whose conservative views he was to uphold throughout the conflicts with the new Republic. His wife, Adèle Chayet, whom he married in 1890, died in childbirth following the birth of their daughter Hélène in 1891. Hélène wrote a biography of her father (H. Duhem 1936), and saw to it that the remaining five volumes of his *Le système du monde* were eventually published in the 1950s.

Following a private primary education, Duhem began his secondary education at the Catholic Collège Stanislas at the age of 11. He took the baccalaureate in both letters (1878) and sciences (1879) but opted for a career in science, influenced especially by one of his teachers, Jules Moutier, who introduced him to thermodynamics. His entrance to the École Normale Supérieure was delayed a year because of acute rheumatoid arthritis, which was to be a recurrent problem. But he

© Springer Nature Switzerland AG 2020
P. Needham, *Getting to Know the World Scientifically*, Synthese Library 423,
https://doi.org/10.1007/978-3-030-40216-7_6

began in 1882 having come first in the entrance examinations. Despite illness and a setback with his thesis, his five years at the École Normale Supérieure were the happiest of his life (Brouzeng 1987, Ch. 1). After three years' study he obtained his degree in 1885.

Earlier, he had submitted a doctoral thesis in physics on 20 October, 1884, which led to his first conflict with those who were to control his career. The thesis was failed by a panel comprising Lippmann (Nobel prize winner, 1908), Hermite and Picard. Although the original documents have not survived, the book (Duhem 1886) published shortly after by the prominent Parisian publishing house Hermann is generally taken to be essentially this first thesis (Jaki 1984, pp. 50–3; Miller 1966, p. 49). He continued to study at the École Normale Supérieure and submitted a thesis on magnetism to the faculty of mathematics in 1888 (Duhem 1888), which was accepted by a panel comprising Bouty, Darboux and Poincaré. The previous year he had taken up a lectureship at Lille, moving to Rennes in 1893, despite success both as a teacher and with his research. He had quarrelled over the treatment of a laboratory assistant with the Dean, who reports that he "almost came to blows with him" (Nye 1986, p. 218; according to Miller op. cit., they actually did come to blows). But Duhem was so dissatisfied with his situation at Rennes that he requested a transfer. He expected a position in Paris, having published three books and some fifty articles. Particularly noteworthy were a book on hydrodynamics (Duhem 1891), important for the understanding of shock waves in air, and his lengthy "Commentaire au principes de la thermodynamique" (1892–94), incorporating the first precise definition of a reversible process and the distinction between quasistatic and reversible processes from Duhem (1887). What he was offered was a chair of physics at Bordeaux, which he considered refusing until friends assured him that it was the road to Paris. The following year, in 1895, the chair was renamed the chair of theoretical physics with the internal appointment of J. A. F. Morisot as professor of experimental physics.

After his unexpected death in 1896, Morisot was succeeded by Gossart, another internal appointment, with whom Duhem didn't see eye to eye at all. Minutes of the Faculty meetings show Gossart complaining that Duhem wouldn't allow him adequate access to instruments locked up under Duhem's jurisdiction, and that he was unfairly disadvantaged in the financing of research costs. Their quarrels continued until Gossart's retirement in 1909, culminating in the machinations leading to the creation of a third chair of physics for Duhem's former student, Marchis. Duhem entered the fray of personal alliances and conflicts which created schisms paralysing the Science Faculty, and saw science at Bordeaux fall behind the other provincial universities. The rector at Bordeaux identified Duhem as the key troublemaker, but as Miller (1966) and Nye (1986, p. 219) point out, rectors were chosen by the education ministry for their reliable republican and anticlerical sympathies. A new rector in 1913 had a different opinion of him, which may have influenced Duhem's election for one of the six newly-created non-resident memberships of the Académie des Sciences late in 1913 (membership of the French Academy being until then restricted to Parisians). Duhem had become a member of several foreign scientific academies and received two honorary degrees abroad, and this belated recognition from his own country came late in his life, when he had only three years to live. Nevertheless, it pleased him enormously, and seems to have inspired him to turn from the life of a recluse to more active social engagement. He remained at Bordeaux until his death from a heart attack at the age of 55 on 14 September, 1916, his hopes of a chair of physics in Paris remaining unfulfilled. He had turned down the offer of a chair in history of science at the Collège de France, saying he would only go to Paris as a physicist. His antirepublicanism led him to turn down the Légion d'Honneur in 1908 because "it would been given by the Republic he hated and signed by a man he detested" (Miller 1966, p. 52).

Duhem was an intransigent figure who might himself be considered to bear some responsibility for bringing his position on himself by his outspokenness and inflexibility. But he received recognition early in his career from abroad, and the lack of appreciation of his ideas by the scientific establishment at home in France reflects more than strictly professional judgements. His opposition to atomistic theories was actually in line with the views of Marcelin Berthelot (1827–1907), a prolific chemist and historian of science who had taken to politics and headed the scientific hierarchy in France at the end of the century. But Duhem had criticised Berthelot in his first thesis, and this is usually thought to be the reason why it was failed. Duhem was in no doubt about Berthelot's hostility. In a letter dated 12 November 1892 to František Wald about the possibilities of publishing a paper of Wald's which takes up Berthelot's principle of maximum work, Duhem says "Several French journals, and in particular the *Annales de Physique et Chimie*, are under Mr. Berthelot's influence. There it is not allowed to raise doubts about or discuss the principle of maximum work" (Pinkava 1987, p. 65). He goes on to say that enquires with Mr. Bouty, editor of the *Journal de Physique*, were more promising. Some three months later, in a letter to Wald dated 3 February 1893, he says

> I would be greatly obliged if you would not publish the letters that I have written to you, and in particular, that concerning Mr. Berthelot and the *Journal de Physique*. The ideas of Gibbs, van 't Hoff and Helmholtz, which I have championed in France, have brought upon me an avowed hostility from Mr. Berthelot which calls for great prudence on my part. As you can judge from my *Introduction à la Mécanique Chimique*, I am not afraid to confront his anger. Nevertheless, publication of the letters I addressed to you some months after the publication of my book could, if it were reported to him, gravely compromise my modest situation. (Pinkava 1987, pp. 68–9)

A projected series of articles by Duhem on the history of science in *Revue des deux Mondes* was curtailed after the first two (Duhem 1895), allegedly at the instigation of Berthelot (Jaki 1984, p. 144), whilst Duhem continued his criticisms in the pages of the *Revue des questions scientifiques* which was beyond Berthelot's sphere of influence. Berthelot's continued resentment ended in 1900, however, after which he voted for Duhem's promotions. But other influential opponents, including Lippmann, le Châtelier and Perrin, ensured that his career as a physicist never took him to Paris. The substance of Duhem's criticism of Berthelot's law of maximum work will be taken up presently, but to put this into perspective it should be noted that Duhem was right (for a modern appraisal of the principle, see Callen 1985, pp.

277–9). Similar criticisms, based on the new thermodynamics, were springboards for the careers of young chemists elsewhere in Europe. Svante Arrhenius, whose thesis had also encountered opposition from his elders, wrote in a review of van 't Hoff's *Etudes de dynamique chimique* (1884) that from its results

> it follows that at normal temperatures Berthelot's "principe du travail maximum" only holds in most cases, and holds for all cases only at absolute zero. Since the reviewer has arrived at the same conclusion, which is supported by an unparalleled amount of experimental data, from completely different grounds, he can only agree completely with the author's view in this case (Arrhenius 1885, p. 365).

6.2   Introduction to Duhem's Philosophy

The "discard and replace" philosophies advocated at various times by Popper, Kuhn and Feyerabend stand in stark contrast to the continuity thesis advocated by Pierre Duhem at the turn of the century. According to the general idea of this thesis, science doesn't progress by leaping ahead and abandoning earlier achievements as a result of rationally resolving conflict or revolutionary takeover, but by gradually building on past gains. However, Duhem didn't model this idea on the simple accumulation of observations and induction by enumeration with which Popper found it convenient to contrast his own view. The picture Duhem paints is far more intricate, as suggested by the tidal metaphor in the passage quoted at the beginning of the chapter, and doesn't issue in a simple formula serving as a criterion for distinguishing good from bad science.

Duhem had very broad interests in science. He was, as Donald Miller says, "that rare, not to say unique, scientist whose contributions to the philosophy of science, the historiography of science, and science itself (in thermodynamics, hydrodynamics, elasticity, and physical chemistry) were of profound importance on a fully professional level in all three disciplines" (Miller 1971, p. 225). His conception of the continuity of scientific progress draws on his research and expertise in all three areas. As a historian he argued against the widely held view amongst nineteenth-century historians of science that there were two great periods of scientific development, that of the ancient Greeks and that covering the seventeenth-century scientific revolution to the present, and there was simply no history to be written about the dark ages of the intervening medieval period. Against this, Duhem argued that there was active pursuit of the critique and development of Aristotelian theory providing antecedents for what are mistakenly taken to be completely new ideas arising in the seventeenth century. Moreover, there is evidence in the writings of seventeenth-century pioneers that can be traced back to medieval sources.

His well-known work in the history of science began in earnest in the first years of the twentieth century. By this time he was well established as professor of theoretical physics at Bordeaux with a number of publications on the philosophy of science from the 1890s which he drew together in his *La théorie physique* (Duhem 1906a [1954])—a book which remains today a classic in the philosophy of science. His interest in the history of science, and particularly the ancient Greeks, is evident in this early work, although he hadn't yet uncovered the medieval texts which were to make his name as a historian. But the influence of his knowledge as a working scientist is what really gives the philosophical work its bite. Perhaps the best-known thesis from this book is often described as the denial that there are decisive experiments in science in connection with his holistic thesis that hypotheses cannot be tested in isolation. This has been misleadingly interpreted to the effect that Duhem denied that scientists could ever make a judgement about whether a hypothesis or theory should be rejected in the light of the evidence. As we will see, that certainly wasn't his view. The thesis is better summarised as Quine succinctly puts it, by saying that "[t]he finality of an experiment is historical, not logical" (1996, p. 163).

Two further general philosophical ideas bear on the continuity thesis. One concerns Duhem's understanding of precision in empirical science, which stands in direct contrast with the conceptions of incommensurability and meaning variance that were proffered by Kuhn and Feyerabend in the 1960s. The second is Duhem's contention that the progress of science involves a striving for unity. The logical positivists who came later were also attracted by the idea of the unity of science, but they interpreted this in terms of a doctrine of reduction. Duhem thought that the physics of his day had been dogged by the metaphysical thesis that all phenomena had to be reduced to mechanics if they were to be adequately explained. His own work on the foundations of thermodynamics led him to turn this thesis around and conceive of unity in terms of incorporating mechanics within the more general theory that he took thermodynamics to be. He was notorious for his opposition to atomism, which in one of its nineteenth-century guises sought to reduce thermodynamics to statistical mechanics. In his 1906 book he construes the drive to reduce all phenomena to mechanics as the latest in a series of preconceived general metaphysical views that he argued had no place in modern empirical science. But again, he didn't oppose this aversion to metaphysics as the logical positivists were to do, on the basis of a verification principle that would restrict the scope and meaning of theoretical concepts to what could be defined in strictly atheoretical, observational or operational terms. His holism was opposed to any such verificationist idea.

The historical thesis of the importance of medieval science for what was to come in the seventeenth century is discussed in the next section. This is followed by sections taking up the philosophical themes indicated above, beginning with the critique of the idea of decisive falsification and followed by a development of the theme. Then we turn to the role of precision in understanding the import of scientific claims. The chapter concludes with a discussion of Duhem's antireductionist conception of the unity of science and some of the difficulties which it raises.

6.3   The Not So Dark Middle Ages

Copernicus published *De Revolutionibus Orbium Coelestium* in the year of his
death, 1543, where he made his revolutionary proposal that the earth is not stationary
at the centre of the universe, but revolves on its axis and moves in an orbit around the
sun. As with Aristarchus, who proposed the heliocentric theory of the solar system
in antiquity, Copernicus' hypothesis of a moving earth was at first rejected. It was
not accepted until well into the seventeenth century, when the so-called scientific
revolution was well under way. There were several reasons for this, a major one
being that Copernicus had no way of countering the arguments that pursuaded the
Greeks that the earth could not be moving.

The Greeks were well aware that observations of the apparent motions of
heavenly bodies establishes only that there is a relative motion between these
bodies and the observer on earth. Consistently with this fact, Aristarchus ("the
first Copernicus") advanced a heliocentric description of the universe with the earth
moving in an orbit around the sun during the third century BC. But there were further
considerations that convinced most Greeks that the earth must be stationary, based
on the kinds of effect they expected to result from such a motion. The daily cycle
of night and day, if not due to the sun's motion around the earth, must be due to
the earth's rotation on its axis. The Greeks estimated the earth's diameter to within
5% of the modern measurement, implying that the circumference at the equator is
approximately 24,000 miles long. A point on the equator would therefore travel this
distance in 24 h if the cycle of night and day were due to the earth's rotation, i.e.
it would move at an enormous speed of 1,000 miles per hour (c. 1600 km. per hr.).
The speed of a point at latitudes north of the equator in Greece would not move
quite so fast, but still at a considerable speed. Airborne objects such as clouds and
birds would be left behind a moving earth, and appear to be moving in the direction
opposite to that of the rotation of the earth at considerable speed. Moreover, a stone
dropped to the ground would not fall to the point directly below the point from which
it was dropped. During the time of descent the point on the earth's surface directly
below the point from which it was dropped would have moved on, and the stone
would lag behind and fall on some other part of the earth's surface (the "lagging
argument"). There is also the problem of centrifugal force. This is something that
can be felt in the tension of a string on the end of which a heavy object is spun
round, and as Aristotle noted, if this object is a bronze cup containing water, "the
water in a cup … [which] is swung round in a circle, is prevented from falling
… although it often finds itself underneath the bronze and it is its nature to move
downwards" (*De Caelo*, II.13). What, then, of all the objects lying loose on the
earth's surface? These would be flung off if the hypothesis of a rotating earth were
correct. But clearly no such effects are observed, and common sense would indicate
that the earth must be stationary and it is therefore that the sphere of the stars must
be moving.

The centrifugal force objection was finally countered by Newton's theory of gravitation in his *Philosophiae naturalis principia mathematica* published in 1687. This theory contains the law of gravitation couched in general principles of mechanics, an important one of which is the law of inertia to the effect that every body continues in its state of rest or uniform rectilinear motion unless acted upon by some external force. The law in this form is generally accredited to Descartes, but the conception of inertia on which it is based was developed by Galileo. In his 1632 *Dialogue Concerning the Two Chief World Systems*, Galileo defends the Copernican hypothesis against the lagging argument with a new interpretation of the observation of freely falling bodies. He introduced the general principle of inertia according to which a body retains its motion when released unless some intervening action prevents it from doing so. A stone released from a tower therefore moves not only under the influence of its weight, drawing it downwards to the centre of the earth, but also under the horizontal motion which the earth's motion imparts to it via the hand of the person who finally releases the stone. This horizontal inertial motion is not directly seen by the observer on the earth, however, because the observer shares it with the stone. Such an observer only sees the respects in which the motion of the falling stone differs from his own, namely the downward fall. The latter point is nothing other than a consequence of the principle of relativity which the Greeks recognised and which Galileo could expect his opponent to accept.

Recognition of the principle of inertia is one of the major new ideas of the seventeenth century which saved Copernicus' theory. Duhem argued that it didn't just appear out of the blue as one of the first contributions to move science forward since ancient times. He pointed to the theory of projectile motion advanced by the influential Parisian philosopher John Buridan (born sometime before 1300 and still living in 1358), who argued that

> after leaving the arm of the thrower, the projectile would be moved by an impetus given to it by the thrower and would continue to be moved as long as the impetus remained stronger than the resistance, and would be of infinite duration were it not diminished and corrupted by a contrary force resisting it or by something inclining it to a contrary motion (quoted by Zupko 2008, section 6, from Buridan's *Subtilissimae Quaestiones super octo Physicorum libros Aristotelis*, XII.9: 73ra).

The idea was already broached in late antiquity by the sixth-century critic of Aristotle's physics and cosmology, John Philoponus, who rejected the Aristotelian conception that what was regarded as non-natural motion of a body can only be sustained as long as a force acts on it. There is no suggestion that Buridan, still less Philoponus, actually had the full and final conception of inertia under a different name. On Philoponus' theory, a projectile once thrown is kept moving until the impetus is consumed, when it follows its natural downward motion. And although Buridan took seriously the possibility that the earth moves, his considered opinion was that it is stationary. The contention is merely that there is a line of development through the middle ages which merits treating impetus as a precursor of inertia. Galileo wasn't the first to criticise the Aristotelian conception of motion, and the idea of inertia that he proposed is clearly reminiscent of these earlier theories. His own conception wasn't exactly the same notion of inertia that was finally established in Newton's first law. In the 1632 *Dialogue* Galileo understood the horizontal motion that the stone retained after release from the tower to be motion in a circle, parallel to the earth's surface, whereas inertial motion in Newtonian mechanics is rectilinear.

This illustrates a general attitude of critical appraisal of Aristotelian doctrine that Duhem argued was fostered within the church, and he draws particular attention to the Condemnation, in 219 articles, issued in 1277 by Étienne Tempier, bishop of Paris. Many of the articles bore on natural philosophy, questioning the necessity of several points in Aristotle's physics if not their actual truth. This prepared the ground for later innovations by encouraging thinkers to countenance the possibility of alternatives to the Aristotelian scheme.

Duhem pursued in detail several aspects to the pre-seventeenth century development of mechanics (Duhem 1903a,b, 1903–1913, 1905–1906, 1906a). But his thesis of the medieval origins of seventeenth century science has been criticised for overemphasising the role of the university of Paris in the middle ages and ignoring Bradwardine and The Oxford "Calculators" from the first half of the fourteenth century. Aristotle thought that the speed of a body was directly proportional to the power impelling it and inversely proportional to the resistance offered by the medium through which it moved. In the extreme case when the density of the medium is reduced to nothing and the resistance is zero, the speed will be beyond any ratio or infinite, which Aristotle thought absurd, leading him to conclude that a vacuum was impossible. This argument was not so persuasive once the idea of impetus had been accepted. His view that when the motive power is less than or equal to resistance, the velocity should be zero is, perhaps, more reasonable. But if speed is proportional to the ratio of power to resistance ($v = p/r$), this makes $v = 1$ and not 0 at equilibrium. And even if the equation is only valid for $p > r$, it entails a discontinuous jump in velocity from 0 to something in excess of 1 as the power overcomes the resistance. Maier (1982) formulates Bradwardine's discovery as imposing the condition that doubling the velocity means squaring the proportion p/r, and in general $n \cdot v \propto (p/r)^n$, which, in modern terms, is satisfied by the solution $v = \log(p/r)$. Bradwardine removed the discontinuity, and was the first to break with the tradition of treating rest and motion as two completely separate phenomena and offer a principle covering both.

Duhem's suggestions of how medieval ideas were actually transferred to renaissance thinkers, for example which texts Galileo actually read in his youth and whether his early manuscripts show that he had read Albert of Saxony, are also criticised. But this is only to further the area of investigation he had opened up, and not to question the fundamental claim of his historical work, that science doesn't progress by radical revolution in which all that has gone before is rejected, but by the continual modification and development of ideas. It is to be distinguished from another line of criticism due to Alexandre Koyré and Anneliese Maier in the decades after Duhem's death, which sought to re-establish the thesis of revolutionary progress, although they placed the revolution somewhat earlier than the seventeenth century. They were the immediate precursors of Thomas Kuhn's *The Structure of Scientific Revolutions* (1962).

6.4 Duhem's Critique of the Idea of a Crucial Experiment

Duhem backed the wrong horses in rejecting Maxwell's electromagnetic theory in favour of Helmholtz's, dismissing (at the start of World War I) Einstein's theory of relativity along with Riemann's non-Euclidean geometry as an aberration of the German mind repugnant to common sense, and was critical of atomic theories of matter. Nevertheless, his philosophical discussion reflects a sensitivity to the rough and tumble of empirical science and has stood the test of time, containing many interesting observations which have gradually been rediscovered in the course of the century. Shortly after the logical positivists broadcast their verification theory, Popper proposed that scientific hypotheses should be distinguished from non-scientific ones by trying to falsify them. Science advances by formulating falsifiable theories, which sooner or later succumb to recalcitrant observations and are replaced by new hypotheses. The problem with metaphysics is not that it is meaningless, but that its claims are not given falsifiable formulations. Some of the shortcomings of Popper's falsificationism were later addressed by Lakatos (1970), who criticised it for its patchwork view and emphasised the integrated, holistic character of scientific theory. These points were not new, but essentially those first made by Duhem in articles from the 1890s and later brought together in his classic 1906 book, translated as *The Aim and Structure of Physical Theories* (Duhem 1954).

Positivism began as a French movement and had taken a firm grip on the French scientific community by the end of the nineteenth century—much to Duhem's displeasure. His heroes were the great French theoreticians of the eighteenth and early nineteenth century such as Lagrange, Laplace, Poisson and Fresnel. But more than that, he thought the ideal of theory-independent observation was a chimera. A popular theme in Duhem's time had it that science proceeds by performing a *crucial experiment*, in which a specific observation decides between two competing theories by falsifying the one and confirming the other. The corpuscular theory of light gave way to the wave theory as a result of Foucault's and Fizeau's experiments, for example. But Duhem questions whether the results of scientific experiments take the form of observations which have the necessary independence of theory to be able to stand alone and play such a decisive role.

Theory and observation cannot be separated into watertight compartments, for "without theory it is impossible to regulate a single instrument or to interpret a single reading" (1954, p. 182). Someone might stumble into a laboratory, see what is going on, and register the "practical facts" of what actions are performed and what is visible to the experimenter. But the significance of what he sees would elude him, and appear as so many unrelated incidents. Hearing the assistant asked to determine whether there is a current in the wire, he will be perplexed to see him looking at the galvanometer. Having learnt this, he will again be surprised to learn that "to the question, 'Is the current on?' my assistant may very well answer: 'the current is on, and yet the magnet has not deviated; the galvanometer shows some defect' ... [b]ecause he has observed that in a voltameter, placed in the same circuit as the galvanometer, bubbles of gas were being released; or else ...", and Duhem gives a long list of possible alternatives (1954, p. 150). If observations of instrument readings provided straightforward observational criteria for the applicability of theoretical statements as the positivists maintained, the theoretical claim that the current is on couldn't be made if the galvanometer didn't display the appropriate reading. But it is not at all absurd to do so, because the theoretical claim is not simply connected to a single observational criterion which the verificationist would take to be its meaning. Anyone conversant with physical theories understands how the theoretical claim is connected in innumerable ways with simple phenomena that an uninitiated person could regard as observable, depending on the circumstances. As Duhem puts it, "it can be translated into concrete facts in an infinity of different ways, *because all these disparate facts admit of the same theoretical interpretation*" (1954, p. 150). It expresses, he continues, "an infinity of possible facts ... in virtue of constant relations among diverse experimental laws ... [which] are ... precisely what everybody calls 'the theory of the electric current' " (1954, p. 151). And it is because the theory is assumed constructed that the theoretical claim is so diversely related to phenomena. There is no question of creating a concise language antecedently to the construction of a theory as the verificationist would have it; "the creation of this language", Duhem concludes, "presupposes the creation of a physical theory" (loc. cit.).

Millikan's measurement of the charge on the electron, based on the determination of the electrostatically induced charge on a number of oil drops by measuring the electric field required to balance the drop against the pull of gravitation and then finding a greatest common multiple, has been discussed from this point of view in Chap. 2.

If experiment were merely the observation of an independent fact, "it would be absurd to bring in corrections" since this would amount to telling a careful and attentive observer that "What you have seen is not what you should have seen; permit me to make some calculations which will teach you what you should have observed" (1954, p. 156). The sceptical reaction that Duhem's words might be expected to evoke is precisely what Galileo encountered when he tried to convince a careful and attentive audience what it was that his pendulum experiments demonstrated. How is observation supposed to confirm the claim that the pendulum's period of oscillation is constant if allowance has to be made for air resistance, friction at the fulcrum and suchlike to which Galileo appealed in order to explain the fact that it finally came to a halt? The pioneers of the scientific revolution had a hard time when faced with the fact that predictions are not confirmed by straightforward observation, but only after apparently devious adjustments have been made. Somehow, the understandable scepticism was overcome and corrections made with a view to accommodating systematic error have become a standard procedure in experimental physics.

The role of correction is to allow the experimenter to form a better picture of what is going on in the concrete instrument of the laboratory—to give "a more faithful reproduction of the truth" (1954, p. 158). Thus, when investigating the compressibility of gases to improve on Boyle's law, Regnault made extensive corrections to the readings obtained in the laboratory. The height of the column of mercury used to record the gas pressure, for example, was corrected to allow for the compressibility of the mercury, itself related to "the most delicate and controversial questions of the theory of elasticity" (1954, p. 147), and for the variations of temperature and atmospheric pressure on the surface of the mercury. A better, if more complex, picture of what is really going on is obtained. As a matter of fact, Regnault is criticised for not realising that a systematic error also arises as a result of the action of the weight of the gas under pressure, and his theoretical picture of the gas under pressure was oversimplified by neglecting the inhomogeneity arising because of the variation of pressure with height. Duhem concludes by describing an experiment in physics as

> the precise observation of phenomena accompanied by an *interpretation* of these phenomena; this interpretation substitutes for the concrete data really gathered by observation abstract and symbolic representations which correspond to them by virtue of the theories admitted by the observer. (1954, p. 147)

Duhem points out that this can and does lead to problems amongst physicists belonging to different schools, or separated by a hundred years, who interpret phenomena differently. Nevertheless, translation can often be achieved. He mentions experiments Newton conducted concerning colours of rings which convinced Newton of the heterogeneous character of light and which he reported and interpreted in terms of his corpuscular theory of light, ascribing different features of reflection and transmission to light corpuscles of various colours. Young and Fresnel reinterpreted these corpuscular features as features of "what they called a wave length" (1954, p. 160), enabling the translation of Newton's results into their wave language. When interpreting scientists immersed in a different culture, Duhem recommends adopting a general stance that Quine and Davidson were later to adopt under the title of the principle of charity, subjecting reports of facts as the physicist of a bygone age saw them to "the usual rules for determining the degree of credibility of the testimony of a man; if ... [he is] recognized as trustworthy—and this would generally be the case, I think—his testimony ought to be received as the expression of a truth". But sometimes it may be necessary to "inquire very carefully into the theories which the physicist regards as established ... seek[ing] to establish a correspondence ... [with] ours, and to interpret anew with the aid of the symbols we use what he interpreted with the aid of the symbols he used" (1954, p. 159). This enables us to make our own assessment of the degree of approximation with which the physicist's testimony supports his interpretation.

There is no trace here of the relativism which Kuhn (1962) was to associate with inter-scientific communication. This is not to say that problems, not to say insuperable difficulties of interpretation, cannot arise in the quest to understand scientists of a different age. Where "authors have neglected to inform us about the methods they used to interpret the facts, ... it is impossible to transpose their interpretations into our theories. They have sealed their ideas in signs to which we lack a key" (Duhem 1954, p. 160).

Attention to the details of accommodating sources of systematic and wildly fluctuating influences has an important consequence for the testing of theories: there can never be a *crucial experiment*. Of course, experimental results may be taken, at a given time and in a given historical context, to argue for one theory and against a competing theory.[1] But Duhem's point is a more general one of principle concerning the bearing of logic on methodology. A hypothesis can never be tested in isolation, but only as one among several theories. As every student knows who has tried to get classical experiments to work, the actual experimental set-up is always considerably more complicated than is suggested by textbook descriptions, which focus only on the leading idea behind the experiment. Whatever the textbooks say, the physicist does not and could not confine himself to the statement he is primarily interested in. When the predicted phenomenon doesn't materialise, what is demonstrated is that the whole body of theory brought to bear in interpreting the experiment contains something which must be rejected. A recalcitrant result doesn't itself determine more specifically, as a matter of logic, where the error lies. The physicist wants to go further and identify a specific source of error, but in so doing he exceeds what the logical rule of *modus tollens* tells him.

Duhem illustrates his case with a discussion of Foucault's experiment to determine whether light travels faster in air than in water, which was at that time widely regarded as a crucial experiment deciding between wave and particle theories of light. According to the corpuscular theory, light should travel faster in water since the refractive index between the two media, being the ratio of the sines of the angle of incidence and the angle of refraction, is on this theory equal to the ratio of the velocity of refracted light to the velocity of the incident light. The value of the refractive index for the air-water interface then implies on the corpuscular theory that light travels faster in water. According to the wave theory, on the other hand, the actual value of the refractive index implies that the direction of the wavefront is changed in such a way as would be explained by light's travelling faster in air. Foucault concluded that the wave theory is correct, and the corpuscular theory is incompatible with the observed facts. Duhem emphasises that the step from the observations to this conclusion is a long one. What Foucault observed with his experiment with rotating mirrors is the appearance of a white and a green spot, the green to the right of the white. Had it fallen to the left, he would presumably have concluded in favour of the corpuscular theory. Clearly, a number of assumptions connect the observations with the system of hypotheses constituting the corpuscular theory and that constituting the wave theory, and the experimental results challenge a whole system of hypotheses and auxillary assumptions, establishing by logic only that the corresponding conjunction of all these hypotheses is false.

But even this may be to ask too much of logic, since "the physicist is never sure he has exhausted all the imaginable assumptions" (1954, p. 190). We now know only too well that there may be ways of developing a particle theory other

---

[1]Duhem often made such judgements. A theory of elasticity due to Fresnel, Cauchy and Poisson "is contradicted by experiment and should be regarded as false" (Duhem 1893, p. 104). Debray's experiments show that "Chemical statics based on kinetic hypotheses is thus condemned" (Duhem 1898, p. 68). And as already mentioned, he persistently argued that Thomsen's and Berthelot's law of maximum work should be rejected (e.g. Duhem 1897; 1902, pp. 156–60). This is discussed again in Sect. 6.7.

than those familiar in the nineteenth century. The discovery of the photoelectric effect and Einstein's theory of this process is usually understood to have suggested a return to the corpuscular idea around the time Duhem was writing. In that case, the import of the corpuscular hypothesis cannot be as definitive as those who took Foucault's experiment to be a crucial experiment supposed. Duhem concluded that a crucial experiment is impossible because the conclusions drawn from experiments are, like theories themselves, tentative and provisional, never immune from the need for readjustment. The fact that, "among the theoretical elements entertained there is always a certain number which the physicists of a certain epoch agree in accepting without test and which they regard as beyond dispute" (1954, p. 211) doesn't detract from the basic principle. "The finality of an experiment", as Quine succinctly put it, "is historical, not logical" (1996, p. 163). Duhem summarises his view as follows:

> The only experimental check on a physical theory which is not illogical consists in comparing the *entire system of the physical theory with the whole group of experimental laws*, and in judging whether the latter is represented by the former in a satisfactory manner. (1954, p. 200)

Duhem criticised some of Poincaré's arguments for conventionalism on the basis of this holistic view. Certain fundamental hypotheses of physics are immune from contradiction by experiment, according to Poincaré, because they are really definitional specifications of the usage of key terms in the physicist's vocabulary. The principle of inertia, for example, is applied in such a way that it cannot be contradicted by experiment. Apparent discrepancies are accounted for by postulating the influence of forces such as friction and gravitation. The argument is developed to reduce all of Newton's laws of mechanics to the status of definitions (Poincaré 1902 [1952], Ch. 6). But Duhem could not see in this a clearly defined principle sharply distinguishing some laws from others. Factors introduced to explain the lack of true inertial movement are of exactly the same kind as the corrections introduced in the application of precise statements of law. A particular theory only provides an accurate account of the experimental situation with the aid of a host of additional assumptions, and these may well include the phenomena of friction and gravitation. Failure of an experiment to agree with predicted behaviour doesn't tell us which of these many assumptions is to be rejected, the predilection of physicists of a certain epoch to accept certain propositions without test as beyond dispute notwithstanding. Fundamental laws forming the cornerstone of whole theories or systems of beliefs are still part of a system which experiment can only show is, as a whole, ill-adjusted to the facts. The conventionalist is correct to say that nothing compels us to reject the cornerstones rather than tampering with other hypotheses instead. Equally, nothing can guarantee that the constant adjusting of the ancillary assumptions and postulating of extraneous causes will not come to be regarded as ad hoc, and an adjustment to a long-cherished principle deemed more satisfactory. "Perhaps someday ... by refusing to invoke causes of error and take recourse to corrections ... and by resolutely carrying out a reform among the propositions declared untouchable by common consent, he will accomplish the work of a genius who opens up a career for a theory" (1954, p. 211). The principle that light travels in straight lines in homogeneous media was maintained for thousands of years. It is the principle we rely upon when checking for a straight line by sight. But the day came when diffraction effects ceased to be explained by the intervention of some cause of error and optics was given a new foundation.

It is sometimes suggested that Duhemian holism unrealistically makes "[the] unit of empirical significance the whole of science", as Quine (1951, p. 42) once put it, leading him to later restrict it to "some middle-sized scrap of theory" (Quine 1960, p. 13) and go on to say that "[l]ittle is gained by saying that the unit is in principle the whole of science, however defensible this claim may be in a legalistic way" (Quine 1975, pp. 315). Other authors say much the same; for example, that "[a]lthough Duhem at times implies that all of science enters as auxillary into any test of a physical hypothesis ..., this is surely an overstatement" (Hoefer and Rosenberg 1994, p. 594). Now Duhem doesn't speak of "the whole of science" but he has a very good reason for not speaking about any specific restriction. His examples often concern the assessment of systematic error, and the restrictions hinted at by these authors would involve a conclusion of such considerations. But as we saw in Chap. 2 and Sect. 4.2, and as Duhem fully appreciated, it is well known that the achievement of stable, reproducible results is not in general a sufficient criterion of adequacy for consideration of systematic error, and there simply is no criterion of completeness for the identification of systematic error. Despite meticulous precautions in experiments on the compressibility of gases, Regnault as we saw was later criticised for having overlooked a source of systematic error: he "neglected the action of weight on the gas under pressure" (Duhem 1954, p. 158). No one before Galileo appreciated the source of systematic error due to twinkling, which disappeared when stars were viewed through a telescope and which resulted in Tycho's measurements of stellar size being scaled down considerably, thereby demolishing one of the arguments against the Copernican theory. Galileo was himself unaware of the systematic error in his discussion of his experiments on motion on an inclined plane arising from failing to distinguish between rolling and gliding down the plane. These are the kinds of example on which Duhem rests his case. Since accommodating the various sources of systematic error might well call upon disparate theoretical considerations which no automatic procedure can finally delimit, it is difficult to see how any "middle-sized scrap of theory" can actually be picked out for the purpose of confining the holistic thesis. It is one thing to say, of course, after the acceptance of a particular set of results within the estimated limits of error, that such and such were the considerations actually brought into play. But it is quite another to delimit the relevant considerations in advance should the results be called into question. Yet this is the position of the scientist faced with recalcitrant experience. What, then, is the force of Quine's latter-day reservations? The claims of science, to which Duhem thought the holism thesis applicable, he characterised as relative and provisional with an eye to their place in the historical development of science, and the problem of delimitation in advance.

6.5 Pushing the Argument Further

Duhem's holistic conception of the empirical import of theories argues against the positivist's strict division between theoretical and observational terms. But like the positivists, Duhem sharply distinguished physics from mathematics, which progresses by adding "new final and indisputable propositions to the final and indisputable propositions it already possessed" (1954, p. 177) with its theorems "demonstrated once and for all" (1954, p. 204). He took a similar stand on matters of definition:

> The mathematical symbols used in theory have meaning only under very definite conditions; to define these symbols is to enumerate these conditions. Theory is forbidden to make use of these signs outside these conditions. Thus, an absolute temperature by definition can be positive only, and by definition the mass of a body is invariable; never will theory in its formulas give a zero or negative value to absolute temperature, and never in its calculations will it make the mass of a given body vary. (Duhem 1954, pp. 205–6)

But science has not confined itself to these conditions. The case of mass, which according to the theory of special relativity varies with velocity relative to the observer, is familiar enough. Negative absolute temperatures have been discussed since the 1950s. Although it is true that no body can have a temperature in virtue of which it would be colder than absolute zero, it turns out that certain bodies might have temperatures in virtue of which they are hotter than any positive temperature, and these are negative temperatures (Ramsey 1956; Braun et al. 2013).

In a famous paper taking issue with two central theses or dogmas of positivism, Quine (1951) adopts Duhemian holism in rejecting what he calls the dogma of reductionism. This dogma, implicit in the verification theory of meaning, that "each statement, taken in isolation from its fellows, can admit of confirmation or infirmation at all" fails because "statements about the external world face the tribunal of sense experience not individually but only as a corporate body" (p. 41). This is essentially Duhem's argument.[2] But Quine extends the scope of the argument to encompass a critique of the analytic-synthetic distinction, which rests on

> a feeling that the truth of a statement is somehow analyzable into a linguistic component and a factual component. The factual component must, if we are empiricists, boil down to a range of confirmatory experiences. In the extreme case where the linguistic component is all that matters, a true statement is analytic. My present suggestion is that it is nonsense, and the root of much nonsense, to speak of a linguistic component and a factual component in the truth of any individual statement. Taken collectively, science has its double dependence upon language and experience; but this duality is not significantly traceable into the statements of science taken one by one. (Quine 1951, pp. 41–2)

Quine arrived at this view after engaging in prolonged discussions with the logical positivists on a "European tour" of the major centres of logical positivism—Prague, Vienna, Warsaw—in the 1930s, and following the subsequent development

---

[2]Which Quine acknowledged in a footnote added to the reprint of his article in the first edition of *From a Logical Point of View.*

of Carnap's ideas. A devastating critique of basing the notion of analyticity on convention is already presented in Quine (1936), where the seeds of much of his later criticisms are also found. Carnap's definition of analyticity rested on a distinction between logical and descriptive truths, the analytical truths being those theorems which remain theorems whatever substitutions are made of the descriptive expressions. Quine counters that the distinction is essentially arbitrary and attempts to justify it are circular.

Quine (1936) argues that if logic is true by convention, this can't be explained as truth by definition since definitions allow only the transformation of truths into other truths; they provide no foundation for them. The explanation given of the alleged conventional truth of the foundations is the adoption of the postulates or axioms, sometimes said to be implicit definitions of the primitive (undefined) symbols. Thus, laying it down that, for example, sentences of the kind "If if p then q then if if q then r then if p then r"—one of Lukasiewicz's axioms for sentential logic—are axioms says something about the meaning of "If ..., then ...". But the supposed adoption of the postulates by convention involves an infinite regress, and could never be carried out even in principle. We have to try to imagine the envisaged conventional stipulation of meaning by postulation to consist, first, of fixing on a list of logical primitives ("if ..., then ...", "not", "every"), considered initially to be meaningless signs, and then running through a list of sentences in which they occur and arbitrarily segregating off some of them and stipulating them to be true. This provides contexts for the initially meaningless marks "If ..., then ...", etc., which are thus conventionally assigned a meaning by rendering the contexts true by convention. If this is to render true all those statements which are, according to ordinary usage, true entirely in virtue of the occurrence of the logical primitives, there must be an infinite number of them. They must, for example, include all the ways of forming a definite sentence from Lukasiewicz's axiom schema by uniformly replacing each of the letters "p", "q" and "r" by English sentences, of which there are infinitely many. The proposed stipulation cannot therefore be made taking one sentence at a time, but must be achieved wholesale in accordance with some general conditions. These general conditions are, as Quine shows in detail, just the laws of logic, so that if logic is to be guided by conventions, then logic is needed for inferring logic from the conventions.

Quine goes on to point out that were the procedure successful, its applicability wouldn't be confined to logic. Should mathematics prove not reducible to logic by definition, the recalcitrant expressions could be accommodated by conventional assignment of truth to various of their contexts in the same manner envisaged for logical truths. Thus mathematics would be rendered conventionally true whether reducible to what is arbitrarily called logic or not. But then there would be no stopping there. The procedure could equally well be extended beyond mathematics to encompass and render conventionally true what is usually thought of as the empirical sciences—the truths of physics, the truths of chemistry, the truths of history, and so on. Any other body of doctrine could be conventionally circumscribed in this way and no body of truths would be characterised in virtue of a special kind of truth or way of being true. Upholders of the analytic-synthetic distinction do, in fact, suppose that there are many analytic statements over and above logical truths. For example, "Nothing is later than itself" and "If Pompey died later than Brutus and Brutus died later than Caesar, then Pompey died later than Caesar" are said to be analytic "in virtue of the meaning of 'later than'". Similarly for the negations of sentences such as "The number two is red" and "The paint growled". If the method could be extended to capture the sentences traditionally regarded as analytic, it could equally be extended much further. Quine's point is that the positivists have provided no criterion distinguishing truths as analytic rather than synthetic.

Quine went on in the "Two Dogmas" paper to suggest that the distinction has never actually been drawn in any general and illuminating fashion. A handful of well-worn examples are often wheeled out as suggestive of a boundary—bachelors are men and vixens are foxes—examples so trite that it is remarkable that they could ever have been thought indicative of any substantial general principle, and certainly not sufficient as an explication of an appropriate general principle. These particular examples hinge on what Putnam (1962) called single-criterion words, whose application is specifiable by a simple formula but which play no significant role in science. As soon as we get on to more interesting classic examples, they have a tendency to become controversial or are now uncontroversially false. Principles associated with causation provide a pertinent illustration. Is contiguity necessary and action at a distance impossible? Do causes always precede their effects? Hume required these principles in his analysis of causation, but they haven't gone without question. Critics haven't found it edifying to hear their challenges confronted with the stubborn declaration that the principles are analytic. Contra Duhem, as we saw, it is now accepted as true that mass can vary with velocity, and absolute temperature can be negative.

Frege thought that in reducing mathematics to logic he would be reducing mathematics to principles which intuition and reflection would show to be indisputable principles. He was extending the method Euclid sought to apply in reducing geometry to self-evident axioms. But when Hilbert (1899) rectified the gaps left by Euclid's axioms, many of his additional postulates were far from simple. It had also been shown that non-Euclidean geometries which conflicted with Euclid's parallels postulate could not be faulted on grounds of consistency. Further complications followed in the wake of Russell's paradox and Gödel's incompleteness results (showing that it is impossible to capture the whole of mathematics in a single, consistent system), and Fregean intuition became irrelevant. What justifies general principles serving as axioms is their consequences; the right axioms are the ones that provide the optimal formulation. Assessing axioms by their adequacy in this sense brings holistic considerations to bear. Justification turns on their effect in systematising the whole system. It can be a complicated and difficult affair, involving the balancing of many factors. But as Duhem emphasises, it is a matter of making a judgement and taking a stand on the issue, not merely adopting a convention with no substantial commitment to truth. To the extent that philosophers pursue ontological and foundational questions, their interests are continuous with science.

6.6 Precision

Duhem thought of his holism as restricted to an area bounded by two extremes. At the one end, the basic principles of logic and mathematics, together with fundamental definitions, stand in splendid isolation as indisputable truths immune from any consequences of their application in sciences. At the opposite end of the scale, there is everyday wisdom as captured in such stalwart truths as "The sun rises in the east" and "Where there's smoke there's fire", which can be relied upon to withstand the onslaughts of scientific rigour. We saw in the last section how Quine's broader conception of holism questions the stance Duhem stubbornly maintained on mathematics, but they seem to have arrived, albeit in their own ways, at a similar position on the second matter. Some years after publishing his "Two Dogmas" paper, Quine restricted holism by introducing a category of observation sentences. This wasn't a return to the positivist's notion of an observation sentence, defined in terms of an antecedently delimited vocabulary of observation predicates. The notion of an observation sentence introduced in Quine (1960) is a weaker one, not presupposing any distinction of predicates as observation or theoretical, but treating it "holophrastically", as an unanalysed unit without structure. In the face of criticism Quine refined his conception in subsequent publications. But since his motives for introducing the notion were more clearly connected with his theory of language learning than the development of science, they will not be discussed further here. In Duhem's account, the notion of scientific precision is central, and although it is connected with experiment and observation, nothing like the notion of an observation sentence is involved. Everyday claims are vague, in contrast with scientific claims which are precise. Scientific precision is intimately connected with the approximate character of scientific claims—something that is at odds with contemporary philosophical usage which associates approximation with vagueness or lack of precision.

Duhem thought that "The sun rises in the east" can be confidently asserted without fear of contradiction because what it does and doesn't assert is so vaguely delimited that it is easily accepted. If, on the other hand, we want a precise law of the motion of the sun, indicating to the observer in Paris what place the sun occupies in the sky at each moment, then, "despite the irregularities of its surface, despite the enormous protuberances it has" the sun will be represented "by a geometrically perfect sphere, ... [whose] position [is] the centre of mass of this ideal sphere" (1954, p. 169). This is where the holistic considerations are brought to bear, Duhem's point being not merely that theoretical considerations enter even simple experiments; it is that they arise particularly in connection with precautions taken to produce stable, reproducible results and reduce error. His comparatively detailed examples illustrate the accommodation of systematic error, bringing to bear laws and theories which seem otherwise quite distant both from the simple considerations in terms of which the experiment was initially conceived and from one another. When observing the position of the sun, the nature of light must be known in some detail to correct for aberration, and used in conjunction with knowledge of the atmosphere to correct for refraction. We saw in Chap. 2 that Millikan's experiment to determine the charge on the electron was based on the idea of comparing a downward gravitational force with an upward electric force. But considerations of systematic error brought the theory of viscosity to bear, leading to a re-evaluation of Stoke's theory, and required a knowledge of the temperature, itself invoking yet new theoretical considerations ensuring that a measure of the volume of mercury in a thermometer provides a reliable measure of temperature.

Approximation enters in connection with the ancillary circumstances. In the case of tracing the sun's trajectory,

> can we not make a single value for the longitude and a single value for the latitude of the sun's center correspond to a definite position of the solar disk, assuming the corrections for aberration and refraction to have been made? Indeed not. The optical power of the instrument used to sight the sun is limited; the diverse operations and readings required of our experiment are of limited sensitivity. Let the solar disk be in such a position that its distance from the next position is small enough, and we shall not be able to perceive the deviation. Admitting that we cannot know the coordinates of a fixed point on the celestial sphere with a precision greater than 1′, it will suffice, in order to determine the position of the sun at a given instant, to know the longitude and latitude of the sun's centre to approximately 1′. Hence, to represent the path of the sun, *despite the fact that it occupies only one position at each instant*, we shall be able to give for each instant not one value alone for the longitude and only one value for the latitude, but an infinity of values for each, except that for a given instant two acceptable values of the longitude or two acceptable values of the latitude will not differ by more than 1′. (Duhem 1954, pp. 169–70; my emphasis)

The law describing the path of the sun cannot, therefore, be upheld in the form of a single formula to the exclusion of the infinity of others that deviate from it within the rationally assessed limits of error. Although "*it would be absurd to say that the same point describes two of these curves at the same time* … [t]he physicist does not have the right to say that any of these laws is true to the exclusion of the others" (Duhem 1954, p. 171; my emphasis).

It has been suggested that what Duhem is arguing here is "that reality must be taken to be as it is revealed to us through observation: fuzzy and imprecise" (Worrall 1982, p. 213). But inferring from the considerations that Duhem raises that reality is somehow fuzzy is gratuitous and finds no support in Duhem's explanations. The italicised parts of the last two quotations clearly show Duhem to be denying any such conception. The point is that the physicist is misrepresented as thinking he is justified in holding a particular function into the real numbers as true. This is not to deny that he holds the truth to be some value falling within specified limits for each of a range of arguments of a specified function, but merely an admission that he doesn't know which. And in order to make the claim as precise as possible, limits of error are specified which are made as narrow as justifiably possible.

Many philosophers primarily concerned to resolve the ancient sorites paradox have a different notion of precision, standing in contrast with the idea of a claim said to hold within certain specified limits of error, which they regard as vague. Duhem's view would be that a claim, such as "The sun rises in the east" characteristically made with no qualifying limits of error, is vague precisely because of the absence of this qualification. The everyday claim is made without the faintest idea of how to go about specifying appropriate limits of error, and this ignorance is manifested by the absence of appropriate qualifications. A scientific claim, on the other hand, is made as precise as possible, bringing to bear a breadth of knowledge as Duhem lucidly illustrates. In general, accuracy increases with the development of the subject, allowing the discernment of detail not previously possible. It was the increased precision of Tycho Brahe's observations that allowed Kepler to distinguish an elliptical form of Mars' orbit from that ascribed by his predecessors.

> Since many of Tycho's determinations of the angular distances between stars were checked by two other observers, it was known that his results could generally be trusted to within two or three minutes of arc. Both Ptolemy and Copernicus speak of a 10-minute error as being tolerable for observational purposes. Without Tycho's reduction of the expected error Kepler's discoveries could not have been made. (Wilson 1972, pp. 93–4)

Ptolemy and Copernicus believed the data available to them justified any claims up to a 10-min limit of approximation, but there is no reason for thinking they couldn't conceive of the observational data being better even if they had no idea how to go about improving it themselves. They certainly didn't think this was grounds for ascribing a degree of fuzziness to reality within the 10-min variation. Who could say where, within the 10-min spread, a more restricted band of values might one day be reasonably held to fall? It would be absurd to say that what Kepler did with Tycho's observations was to falsify Copernicus' account of how fuzzy the world is. Rather, he constructed a curve which fitted within a narrower margin of error, showing how the Copernican apparatus of circles on circles could be replaced with a simpler description and paving the way for Newton, which led in turn to the finer distinctions improving Kepler's own theory (Needham 1998, 2000).

6.7 Reduction, Unity and Progress

When he argued against the idea of progress on the basis of the subsumption of theories by later theory, Feyerabend (1962) overlooked the point about the approximate character of scientific claims. He gives the example of Galileo's law of free fall (that freely falling bodies on the surface of the earth fall with a constant acceleration), which on the basis of his "meaning variance" thesis, according to which the meaning of terms in a theory is determined by the general principles of that theory, is inconsistent with Newtonian physics. For Newtonian physics implies that the acceleration increases as a freely falling body falls. This increase is minuscule over the distance of fall "at the surface of the earth" compared with the distance from the surface to the centre which determines the gravitational force on the body. The Duhemian point would be that it is unreasonable of Feyerabend to interpret Galileo's claim as implying distinctions within his estimated limits of experimental error.

Duhem makes precisely this point in reference to the familiar claim that Newton generalised, and so incorporated, Kepler's laws in his theory.

> The principle of universal gravity, very far from being derivable by generalization and induction from the observable laws of Kepler, formally contradicts these laws. If Newton's theory is correct, Kepler's laws are necessarily false. (Duhem 1954, p. 193; emphasis in original)

For according to Newton's theory, each planet moves not only under the influence of mutual gravitational attraction with the sun, which would result in elliptical orbits, but also under the influence of all the remaining mass in the universe, especially the other planets, which distort the regular elliptic path. Duhem argues that Kepler's laws are, according to the criterion established by Newton's laws, only approximately true, and whatever certainty there is in Newton's theory does not emanate from Kepler's laws. What support they have arises, rather, as a result of calculating

> with all the high degree of approximation that the constantly perfected methods of algebra involve, the perturbations which at each instant remove every heavenly body from the orbit assigned to it by Kepler's laws. [T]hen ... the calculated perturbations are compared with the perturbations observed by means of the most precise instruments and the most scrupulous methods. Such comparison will not only bear on this or that part of the Newtonian principle, but will involve all its parts at the same time; with those it will call in the aid of all the propositions of optics, the statics of gases, and the theory of heat, which are necessary to justify the properties of telescopes in their construction, regulation and correction, and in the elimination of errors caused by diurnal or annual aberration and by atmospheric refraction. It is no longer a matter of taking, one by one, laws justified by observation, and raising each of them to the rank of a principle; it is a matter of comparing the corollaries of a whole group of hypotheses to a whole group of facts. (Duhem 1954, pp. 193–4)

As we saw in the last section, Kepler was a scientist, well aware of matters of observational error, and is not reasonably interpreted as holding a claim unqualified by justifiable estimates of error.

The need to consider "a whole group of hypotheses", rather than isolated claims, when assessing whatever truth can reasonably be ascribed to them is the central point of Duhem's holism. A particular hypothesis cannot be considered independently of those factors delimiting the degree of approximation with which it purports to be upheld at a given point in history. He argued against Poincaré that in interpreting formulas as statements of physical theory, we shouldn't draw consequences by importing to them a more definite determination than experiment can support. Moreover, in confining interpretation within these limits, we must take into account all the considerations which lead to the interpretation which observation does sustain. Such considerations are drawn from the whole body of knowledge at the scientist's command; they are not easily delimited and enumerated, but cannot reasonably be ignored. Accordingly, reduction should be formulated so as to accommodate facts such as that although Galileo's law is not strictly speaking in agreement with Newton's laws (which require a slight change in acceleration as terrestrial objects fall, whereas Galileo's law specifies constant acceleration), the limits of observational accuracy attainable in the seventeenth century allowed no detectable distinction to be made. Here we should say that what is deduced or derived, though different, would not have been distinguishable from the historically faithful formulation, all things considered. The idea that earlier and later theories are separated by an abyss which can't be bridged by rational argument, making the views of the respective theorists *incommensurable* and rendering impossible a rational evaluation of which is better, is therefore excessive. A historical overview of scientific progress would be inconceivable. This is not to say that interpreting claims from the past is without problem, as we saw in Sect. 6.3. But difficulties of this sort do not constitute objections to the general claim that newer theories subsume older ones by approximation. The claim is not that it is easy to interpret past science.

Just as Duhem hoped to overcome diachronic barriers to understanding, so he envisaged the overcoming of synchronic barriers between the various areas of scientific study in a unified science, at least as an ideal towards which science does and should strive. The positivists had a similar ambition, but thought that unification was to be realised by the ultimate reduction of all sciences to a single, fundamental science for which they look to physics. This view continued to enjoy support from many philosophers long after positivism fell out of favour, but has met with increasing resistance in recent years. It was never Duhem's conception of what would be involved in unity. He thought that as science progresses by continually retouching and modifying its theories to deal ever more successfully with recalcitrant cases, it would expand and develop new concepts in rigorously formulated principles which could be systematically elaborated. He put his ideas into practice in his scientific work contributing to the early development of physical chemistry.

Duhem was opposed to the prevailing opinion of his day that progress and understanding in physics should be pursued by reducing phenomena to mechanics, and championed the view that thermodynamics should be understood as incorporating mechanics rather than being reduced to it:

> ... If the science of motion ceases to be the first of the physical sciences in logical order, and becomes just a particular case of a more general science including in its formulas all the changes of bodies, the temptation will be less, we think, to reduce all physical phenomena to the study of motion. It will be better understood that change of position in space is not a more simple change than change of temperature or of any other physical quality. It will then be easier to get away from what has hitherto been the most dangerous stumbling block of theoretical physics, the search for a mechanical explanation of the universe. (Duhem 1894, pp. 284–5)

Put another way, mechanics is expanded by the addition of concepts and principles from thermodynamics, and the whole new body of knowledge is integrated in a systematic new theory. All the old mechanics is incorporated, but in addition there is provision for the explanation of new phenomena. This includes phenomena involving temperature, heat and chemical reactions which couldn't even be expressed within the mechanical framework, but also mechanical effects requiring non-mechanical explanations. For example, pressure is a typical mechanical feature, but cannot be explained purely mechanically when arising as osmotic pressure, which calls for a distinction of chemical substances. Whereas mechanics only recognises mass as the measure of the amount of matter in a body, thermodynamics divides the mass into amounts of different substances making up the body, and treats energy not simply as a function of mass, but as a function of the amounts of the different substances making up the mass of a body. This provides the materials for a thermodynamic explanation of osmotic pressure.

Although opposed to reduction, Duhem rallied against the barrier between physics and chemistry that still reigned amongst chemists in his day. In the nineteenth century, chemists considered properties like melting and boiling points to be physical properties which could be used to distinguish substances (for example, isomers produced in a particular reaction) and perhaps even identify them (provide a general necessary and sufficient condition for being the substance in question). But these were regarded as chemical epiphenomena not essentially connected with the reactivity of a given substance with other substances, which was regarded as the specific domain of chemistry. Marcelin Berthelot appealed to this distinction in defence of the principle of maximum work, which had been proposed in the middle of the century by Julius Thomsen. In his first book, Duhem formulates the principle as follows:

> All chemical action accomplished without intervention of extraneous energy tends towards the production of the system of substances which release the greatest heat. This principle entails the following consequence: of two conceivable reactions, the one the inverse of the other, and releasing heat while the other absorbs heat, the first alone is possible.
>
> The heat released by one reaction without bringing into play any external work is the decrease to which the *internal energy* of the system is subject by the effect of the modification. Consequently, in accordance with the rule proposed by Berthelot, the possibility of a reaction presupposes that this reaction produces a decrease in energy. The stability of a chemical equilibrium is therefore assured if the equilibrium state corresponds to the smallest value that the energy of the system can take. In a word, this rule has it that energy plays the role in chemical statics that the potential plays in statics proper. (Duhem 1886, p. ii)

Although there are many cases which conform with the principle, there are exceptions. It could only be saved by what Duhem held to be an illegitimate demarcation between chemical phenomena, alone subject to the law, and changes of physical state which Berthelot held to be exempt from the law.

> Sulphuric acid, for example, combines with ice and this combination produces cold. In order to bring this exception within the rule, the reaction must be divided into two phases: one part being the fusion of ice, a *physical* phenomenon which absorbs heat, and the other part, the combination of liquid water with sulphuric acid, a *chemical* phenomenon which releases heat. But it is by a purely mental conception, and not as a representation of reality, that it is possible to thus decompose a phenomenon into several others. Moreover, accepting that chemical phenomena obey the law of maximum work while physical changes of state would be free is to suppose that there is between the mechanism of these two orders of phenomena a line of demarcation which the work of Henri Sainte-Claire-Deville has removed. (Duhem 1886, pp. ii–iii)

The question was squarely placed within the general issue of the unity of science, and Duhem argues that Berthelot's appeal to the distinction between physical and chemical processes in defence of the principle of maximum work is ad hoc.

Berthelot's rule supposes that a chemical reaction produces a reduction in internal energy of the reacting material, and thus that a stable state of chemical equilibrium corresponds to the lowest possible value of energy of the system, just as does the stable state of a mechanical system. The failure of Berthelot's rule shows that energy alone cannot serve as the criterion of chemical equilibrium, and if the analogy with mechanical systems is to be upheld, a generalisation of mechanics is required. Something other than energy must be found to play the role analogous to that which the potential plays in mechanics. Duhem goes on in his Introduction to the 1886 book to show how work in thermodynamics by Massieu, Horstmann, Helmholtz and Gibbs had led them to a better appreciation of the conditions governing chemical equilibrium, and proceeds to further develop the thermodynamic potentials, which were adequate to cover all cases of chemical equilibrium without ad hoc distinctions, in the body of his book (Needham 2008a).

This conception of unification by expansion and systematic integration can be applied to other areas, even where Duhem himself opposed such development. He was notoriously opposed to atomic theories. But although the reality of the microstructure of matter is now accepted, the reductive thesis that macroscopic phenomena can be reduced to microscopic phenomena no longer carries the unstinting support it once enjoyed in the mid twentieth century. Sklar (1993, pp. 367–73), for example, suggests that the only motivation for the additional hypotheses needed to complete Boltzman's plan of reducing thermodynamics to statistical mechanics is the anticipated reduction. The relation between thermodynamics and statistical mechanics is not one of reduction but rather a case of the one complementing the other in an integrated theory. From the Duhemian perspective, there is no preconceived notion of "fundamental law" or "basic property". Microscopic principles complement macroscopic theory in an integrated whole, with no presumption of primacy of the one over the other. From the long historical perspective, the body of systematic theory has been seen to grow. If, as seems reasonable, it will continue to do so in the future, who knows what additions may accrue? There is no telling whether, or to what extent, Duhem's dream of a unified science will be realised. Perhaps it just won't prove possible to bring the various threads together into a consistent, systematically unified whole. We might introduce the term "physical property" for whatever properties feature in this future science, be it unified or not. But there is no telling what the term "physical property" might or might not eventually cover. The so-called closure principle favoured by latter-day reductionists has no real force. As one formulation of the claim has it, "If a physical event has a cause at t, it has a physical cause that occurs at t" (Kim 2005, p. 43). But this is at best an empty tautology, to the effect that whatever eventually yields to systematic theory eventually yields to systematic theory.

Chapter 7
Realism and the Advancement of Knowledge

7.1 Cognitive Progress in Science

Niiniluoto has finessed the Popperian understanding of scientific progress as increase in the experimentally verified degree of verisimilitude. But taking as he did the extraordinary growth of scientific knowledge as a major challenge facing the philosophy of science, Popper clearly understood scientific progress to consist in the growth of knowledge. Duhem was with Popper on this point, although he conceived the advance of science as a steady, if halting, successive refinement and improvement of knowledge. It has been emphasised here in Duhemian spirit that a typical scientific claim includes an explicit estimate of the probable error and presupposes an analysis of the systematic error, the claim being to know that the specified result lies within the specified limits of error under the presupposition that the conditions adumbrated in the account of systematic error hold good. Theories are accepted as accommodating the results of experiments and observations within the specified limits of error. Unless these assessments are retracted, for example as the result of a re-evaluation of the sources of systematic error, theories once accepted are usually preserved as limiting cases adequate for the intended domain and within the declared limits of error. Of course, knowledge claims made at one time are sometimes retracted in the light of knowledge acquired later. This is all part and parcel of Duhem's continuity thesis, which is not the simplistic accumulation of observations with which Popper contrasted his views. The deeper claims about the nature of reality—what really exists (ontology) and drives change or underlies stability—evolve with the development of more precise theories. Occasionally, claims that were taken as well established are rejected. Nevertheless, the steady progress at issue in Duhem's thesis concerns the generally accepted scientific consensus emerging once controversies were settled and uncertainties that didn't command general acceptance were resolved. Or so a realist view of science would have it. But this view has not gone unchallenged.

© Springer Nature Switzerland AG 2020

P. Needham, *Getting to Know the World Scientifically*, Synthese Library 423,

https://doi.org/10.1007/978-3-030-40216-7_7

Kuhn and Laudan were critical of realist attitudes towards science. Kuhn rejected a correspondence theory of truth, arguing "There is ... no theory-independent way to reconstruct phrases like 'really there'; the notion of a match between the ontology of a theory and its 'real' counterpart in nature now seems to me illusive in principle" (Kuhn 1970, p. 206). This passage has been quoted before, in Sect. 3.2, where the point was made that knowledge of the truth of a theory doesn't require theory-neutral judgements in the form of independent access to the theory and what it describes. We have seen how theory-ladenness is built into the cumulative conception of scientific progress as Duhem describes it. Laudan (1981) espoused a pessimistic induction, according to which the many past failures of science in the form of discredited theories, such as Descartes' vortex theory of the solar system and the phlogiston theory of combustion, lend strong inductive support for the conclusion that present science is similarly wrong. Kuhn and Laudan offer an account of scientific progress based on problem solving amounting, crudely put, to an increase in the number of problems facing the current paradigm that are solved by a new paradigm, even if this is at the expense of some solutions provided by the old paradigm not addressed by the new paradigm ("Kuhn loss"). For example, Descartes' vortex theory proffered an explanation of why the planetary orbits lie in the same plane, whereas Newton's theory of gravitation which replaced it left this unexplained but explained, or explained more satisfactorily, many more phenomena, such as terrestrial projectile and free fall motion, tidal motion and detailed lunar and planetary orbitals.

On the other hand, it could well be said that solving problems provides evidence for the theories and ancillary assumptions yielding the solution, thus contributing to the accumulation of knowledge. In the light of what we presently take to be knowledge, past successes may sometimes be evaluated as incorrect. Explanations of the planetary orbits lying in the same plane are now based on certain initial conditions—the formation of a cloud of material flattening out and extending over the range covered by the planets by the forces connected with the circular motion around the sun. The Kuhn loses may thus be firmly rejected, when they should no longer be weighted against the merits of the theory whose adoption led to their rejection as the problem-solving criterion of progress would have it.

This is not to imply that scientific knowledge is always gained as a result of problem solving. Much knowledge is simply the result of persistent observation and data collection. Think of the enormous body of knowledge accumulated by naturalists that Darwin and Wallace were able to build on, and which biologists continue to rely upon. Much of chemists' time is spent ascertaining the standard properties of materials and their dependence on ambient conditions. Captain James Cook's voyages of exploration enabled him to draw up maps which have not been faulted since but only modified for greater accuracy. Astronomers have trawled the night sky to draw up stellar atlases, and so on, without necessarily producing dramatic solutions to puzzles. The significance of the more routine exercises in laboratory practice and strictly observational procedures is not to be compared with trivial truth not worth pursuing. But it might be thought an inferior or primitive form of knowledge insofar as it falls short of providing understanding. Searching for solutions to problems might well be cast in the form of a quest for understanding. Understanding is itself knowledge—knowledge of how or why something is the case, perhaps in the form of what caused such-and-such or which laws provide a general insight into the phenomenon—and a contribution to understanding the world is a particularly significant contribution to scientific progress. Other contributions may make more modest contributions to progress. But making little progress shouldn't be confused with making no contribution to progress at all or merely promoting without actually making any progress.[1]

The Kuhn loss in the case of Descartes' proposed explanation of the planetary orbits lying in the same plane might be considered a loss in our understanding of the world. It is important to bear in mind the scientific standing of the proposal, however. Descartes' theory was part of a controversial new view of the universe that was by no means generally accepted by the scientific community at large at the time. It offered a possible explanation of a phenomenon for which no other explanation was currently on offer, which accounts for the interest in this case. But a possible explanation based on inaccurate or false claims is not an explanation and so hardly a real loss. However interesting, it is a suggestion that proved untenable. But not everyone agrees that explanations must be based on what are taken to be facts.

Finnur Dellsén has recently suggested that explanations yielding understanding that counts as scientific progress doesn't necessarily do so.[2] In 1905, Einstein proffered a statistical mechanical explanation of Brownian motion—the random motion of small but visible particles suspended in a liquid medium that had puzzled scientists for decades. But Einstein was cautious:

> It is possible that the movements to be discussed here are identical with the so-called Brownian molecular motion; however, the information available to me regarding the latter is so lacking in precision, that I can form no judgment in the matter. (Einstein 1905 [1956], p. 1)

Some years later Perrin conducted his famous experiments on Brownian motion and confirmed Einstein's theory. In this connection, Dellsén's claim, that "Einstein clearly gave us understanding" (2018, p. 455), is ambiguous. Einstein held his account unjustified in 1905, but Perrin's later work did justify it (and in this respect it differs from Descartes' proffered explanation of the planets lying in the same plane). So it is reasonable to claim that Einstein gave us understanding after Perrin's work, but prior to that we could only say that Einstein put forward a proposal that

---

[1]This point is sometimes obscured in the literature by contrasting fruitful explanations with extremely trivial gains in knowledge, such as counting grains of sand in a particular location or establishing spurious correlations such as increases in birth rates and local stork populations. Even so, minuscule, not to say negligible, advances are just that, hardly worth the effort of acquiring them perhaps but not counterexamples to the thesis that progress is epistemological progress.

[2]Dellsén criticises Bird's (2007) defence of the thesis that scientific progress is the advancement of knowledge by "assum[ing] that it is possible for there to be increases in scientific understanding without accumulations of scientific knowledge" together with claims about what is "most natural" (Dellsén 2016, pp. 74, 80). It is as well to point out that participants in this discussion fall back on conflicting intuitions. In the following paragraphs I give vent to some of mine.

would explain Brownian motion, however cogent the account appeared, *were* it to be empirically vindicated.[3] Unlike many other hypothetical explanations offered over time that never became more than hypothetical and therefore didn't constitute scientific progress, Einstein's theory was borne out by the evidence. So although it could not be definitively deemed progressive in 1905, it did eventually count as making a contribution to progress.

It is apposite to point out that understanding is not infrequently acquired by the persistent accumulation of knowledge in small stages, each not itself amounting to a remarkable advance in understanding. An example of the gradual elucidation of an important concept resulting from the massive investigation (with at one point a paper published every hour) of relevant phenomena is the search for an understanding of hydrogen bonding (see Needham (2013) for an overview). This is typical of the acquisition of knowledge in many branches of science, to which the more spectacular advances shouldn't blind us.

Dellsén also appeals to explanation in terms of minimal idealisations (Weisberg 2007)—models encapsulating core causal factors in a simplified context ignoring or overemphasising other factors which otherwise would make the model complex and intractable—which overlook "explanatorily irrelevant factors" (Dellsén 2016, p. 81) by neglecting resistance, assuming infinite populations, neglecting intermolecular collisions, and so on. But is it surely an oversimplification to say that such models "are not true" (Dellsén 2016, p. 81). The relevant factors are approximately truly described by such models, and are accepted as approximate explanations, holding within the limits of precision of the approximately described relevant factors and with the proviso that the misrepresented factors do not come significantly (within the limits of precision of the model) into play. The kinetic model neglecting collisions of the ideal gas law holds in this sense under low pressure but is not applicable to the behaviour of gases under high pressure when intermolecular collisions cannot be ignored.

There are many ways in which science may be said to progress: by increased funding, larger numbers of active scientists, increased technical skills among researchers, an expanding pool of relevant speculations, and so on, much of which contributes to eventually establishing cognitive progress with the securing of well-founded claims. But even in the specifically cognitive sense of scientific progress, in the theoretical representation of the world by science of primary concern here, it covers the furthering of knowledge across the board and is not a monolithic concept confined to furthering understanding.

---

[3]Something like this is what I take to be the main thrust of Park's (2017) critique of Dellsén, although I wouldn't accept his ideas about means for increasing scientific knowledge (Dellsén 2018, pp. 452–3). I should add that the "would ... were it empirically justified" qualification of "explain" is often tacitly understood and not explicitly mentioned. But it is readily made explicit in contexts where, like those considered here, we want to claim that Descartes' proffered explanation of the planets lying in the same plane doesn't give us understanding whereas Einstein's statistical mechanical explanation does.

Antirealists maintain that we are not justified in holding that the theoretical representation afforded by science is in any sense a true representation of how the world really is. Laudan's pessimistic induction mentioned above is one such argument to this effect, complementing Kuhn's argument from the putative incommensurability of successive theories and an argument not yet mentioned by Bas van Fraassen from the so-called underdetermination of theory to the same general effect. These arguments are challenged in the following sections, but without going to the diametrically opposed extremes advocated by some realists discussed in succeeding sections. In the course of this discussion I try to steer a middle path of a moderate realism that acknowledges the force of the natural view that science is to be taken literally, accommodating change over the course of history and recognising that there are aspects of theory that scientists themselves don't take literally.

7.2 Superscientific Inference

Laudan's pessimistic induction is what might be called a superscientific inference, or meta-inference, being about scientific practice and inference. In this respect it is like Putnam's "miracle argument", advanced by realists as an argument for the diametrically opposed conclusion that science is coming ever closer to the truth. This should already raise our suspicions that such appeals to one and the same history of scientific practice are insufficient for drawing either conclusion. I will take up the "miracle argument" after considering the plausibility of the pessimistic induction in a little more detail.

The many past failures of science, so the pessimistic induction runs, strongly supports the conclusion that present science will also be shown to fail. The contrary conclusion drawn by the realist is based on a fallacy:

> the fallacy of affirming the consequent[4] is indeed fallacious. When W. Sextus or R. Bellarmine or Hume doubted that certain theories which saved the phenomena were warrantable as true, their doubts were based on a belief that the exhibition that a theory had some true consequences left entirely open the truth-status of the theory. Indeed, many nonrealists have been nonrealists precisely because they believed that false theories, as well as true ones, could have true consequences. (Laudan 1984, p. 242)

Laudan supports his claim by enumerating a list of theories and theoretical entities from the history of science (which "could be extended ad nauseam" 1984, p. 231) that had their advocates in their day but have since been discarded and whose central terms are now deemed not to refer to anything: the crystalline spheres of ancient and medieval astronomy, the theory of circular inertia, the humoral theory of medicine, the effluvial theory of static electricity, the phlogiston theory of chemistry,

---

[4][I.e., inferring "p" from "If p then q" and "q", or inferring "a is A" from "All As are B" and "a is B". For example, inferring "Smith is a communist" from "All communists value freedom of speech" and "Smith values freedom of speech".]

the caloric theory of heat, the electromagnetic ether, etc. These theories make false existential and associated claims, for example in the case of the caloric theory that heat is conserved, which was understood in terms of a persisting entity. The immediate conclusion is that the same goes, in all likelihood, for our present theories too. A corollary, that only wishful thinking leads anyone to believe that "realism alone explains why science works" (p. 245), has something to be said for it, although not for the reason that Laudan gives, namely the pessimistic induction.

Although many past scientific claims have been subsequently rejected, their definitive falsification is itself a contribution to knowledge often embodied as an entailment of some presently accepted theory. For example, thermodynamics implies that the defining condition of caloric, that heat is conserved, is false, while preserving the insights and distinctions underlying the notions of specific and latent heat for which the caloric theory was first invented. But this is by no means the dominating picture. Many important and long-standing discoveries have never been refuted. J. W. Gibbs's groundbreaking work "On the Equilibrium of Heterogeneous Substances" (1876–1878), for example, which formed the theoretical foundation of physical chemistry at the end of the nineteenth century, "has required no correction since it was published . . . The underlying principles are few, and rigorous" (Sengers 2002, p. 43). If we also take into account experimental error which has always been incorporated into good scientific claims, then progress is made by improving on older procedures as illustrated by Kepler's discoveries in the last chapter. In surpassing the older results by more precise ones, the older results, taken with the associated limits of error, are not really falsified. On both counts, then, we see new theories avoiding pitfalls of rejected predecessors and incorporating their successes. Viewing history in this light, it would not be an appropriate induction to infer that a false successor theory will in turn be followed by another false theory rather than conclude that we're improving our knowledge and getting closer to the truth. There is no good reason why the pessimistic pattern should be imposed on the history of science and construed as the uniformity which should be projected into the future.

It might be rejoined that Laudan is right about the deeper ontological claim of the eighteenth-century theories of heat, chemical combination and the state of substances, namely the existence of caloric, which was overturned by subsequent theory. In this spirit, fellow antirealist Kyle Stanford claims that "the caloric theory . . . illustrate[s an] . . . important way that even a highly successful theory might fail to be even approximately true in its claims about otherwise inaccessible domains of nature" (2018, p. 216). But this overplays the status of caloric, which was not so firmly entrenched as this story would have us believe. Joseph Black, who was perhaps the first to moot the idea in connection with his notions of specific and latent heat, was notoriously reticent about it. And Lavoisier, to whom we owe the term "caloric", didn't hide his doubts about the adequacy of the notion:

> It is by no means difficult to perceive that this elasticity [of air] depends upon that of caloric, which seems to be the most eminently elastic body in nature. Nothing is more readily conceived, than that one body should become elastic by entering into combination with another body possessed of that quality. We must allow that this is only an explanation of elasticity, by an assumption of elasticity, and that we thus only remove the difficulty one

step farther, and that the nature of elasticity, and the reason for caloric being elastic, remains still unexplained (Lavoisier 1789, p. 22).

Fellow chemist Humphrey Davy (1778–1829) held that the material theory of caloric was one of two major defects of Lavoisier's theory, alongside the neglect of light—which he held was matter, and oxygen gas was a compound of base of oxygen and light (Knight 1978, p. 33). He generated heat by rubbing and melting ice, supporting Benjamin Thompson's competing theory of heat as motion of the microscopic constituents of bodies, and physicists wrangled over the pros and cons of each theory. There was no consensus, and the downfall of caloric was not the overturning of a generally accepted theory. They did have a common failing, which was corrected by thermodynamics. As Duhem pointed out, although

> physicists were divided regarding the nature of heat, they were unanimous in recognising that the heat absorbed by a body during a transformation depends only on the state from which the body departed and that to which it arrived. ... [But it proved] necessary to recount ... the history of the body from the start of the change until the end ... for, if it is brought back to this [initial] state after having undergone a series of changes, it will in general have absorbed more heat than it has released, or released more heat than it has absorbed. It is still possible to speak of the quantity of heat absorbed or released by a body while it is undergoing a determinate transformation. But it is not any longer possible to speak of the quantity of heat enclosed in a body taken in a given state. (Duhem 1895, pp. 403–4)

But this common failing itself hardly amounted to a theory.

A similar tale might be told about the crystalline spheres of the Aristotelian cosmology, which agreed with the simple qualitative picture of the geocentric system with the planets and sun moving in circular paths about the earth but not with the complex systems of deferents and epicycles that were called upon to give a quantitive account of the planetary motions. An antirealist view was conjured up to accommodate this discrepancy (see Sect. 7.4 below), but whether this satisfied realists in general and those conversant with the technical details, as distinct from Church dogmatists, is another matter. The Greeks were in the habit of speculating about alternative possible hypotheses to accommodate the phenomena, as illustrated several times in this book, without claiming conclusive support from the evidence. But Laudan makes no allowance for tentative suggestions. Whether it is good scientific argument for us to generalise from stances taken in ancient and medieval times which weren't regarded then, and which we wouldn't regard, as firmly based on the evidence then available is doubtful. The "deeper" aspects of theories concern the development and refinement of concepts in the formulation of general principles, and ontological claims are governed by the requirements of these principles rather than the more speculative and controversial ideas on which Laudan focuses.[5]

---

[5] The pessimistic induction has been roundly criticised as a bad argument by many philosophers (see Mizrahi 2013 and references cited there). Norton (2003, 2006, 2010, 2014, 2015) goes further, arguing that inductive inference as actually practised in science is not plausibly construed as based, like deductive inference, on general principles such as induction by enumeration to which Laudan appeals. Rather, it takes the form of what he calls material inductive inference, relying on specific factual premises holding in restricted domains and therefore applying locally rather than generally.

Kyle Stanford has more recently sought support from the historical record for an argument to much the same conclusion as the pessimistic induction by appealing to *unconceived alternatives*. But as we saw with the caloric theory, his historical cases also focus on the more speculative ideas rather than the more substantial gains that came to be accepted by the scientific community at large once uncertainties were cleared up and controversies had settled, i.e. gains of the sort at issue in Duhem's continuity thesis. No detailed exposition of his presentation of his argument is given here, although I sometimes note how this point about his historical focus bears on the cases considered below. But there is an important methodological point worth bringing to the fore. When selecting theories as "not even approximately true", or "fundamentally distinct" from another, or as being "profound and fundamental", his strategy is "not to seek to provide definitions or necessary and sufficient conditions, but instead to flesh out our understanding … using a range of exemplars drawn from the historical record of scientific inquiry itself" (Stanford 2018, p. 215). Much the same could be said of the terms Kuhn used to describe his notion of paradigm shift, scientific revolution and conceptual change, which as we saw were hardly clarified by Wray's suggestions (Sect. 3.2). The argument hangs on the perceived force of the examples and the interpretation of their role in the historical development of science.

By contrast with Laudan's antirealism, it is often remarked that there is a strong realist streak in many contemporary scientists. As a preliminary to Putnam's "miracle argument", let us note an expression of this. With nineteenth century doubts about the existence of molecules still a lively memory, the French chemist Robert Lespieau argued

> in a book entitled *La molécule chimique* (1920), … that we are justified in seeing in the formulas currently in use "an image of reality," although it is only "a very crude image." For, he asked, if they had nothing to do with reality, wouldn't it be strange that such images result in so many successful predictions? (Nye 1993, p. 147)

Here Lespieau is arguing for the existence of molecules. Putnam advances a seemingly analogous argument for a much grander thesis, that it would be not merely strange, but miraculous if the consistently successful predictions of science were not to be explained by its being at least approximately true.

> The positive argument for realism is that it is the only philosophy that doesn't make the success of science a miracle. That terms in mature scientific theories typically refer (this formulation is due to Richard Boyd), that the theories accepted in a mature science are typically approximately true, that the same term can refer to the same thing even when it

---

It would be a mistake, for example, to infer from the fact that some quantities of matter have sharp melting points that all do. With its pretensions to generality, Laudan's pessimistic induction could not have a specific factual premise that would justify its conclusion. Several authors (Dorling 1973; Norton 1993, 1994) have argued in recent years for the importance of a deductive route to scientific discovery via the elimination of alternatives without a sceptical Popperian twist, justifying belief on the basis of the evidence in claims susceptible to critique, modification and perhaps outright rejection in the light of further evidence. Demonstrative induction is a special limiting case of Norton's material induction where "the material postulate and the evidence taken together deductively entail the hypothesis at issue" (Norton 2003, p. 665).

occurs in different theories … these statements are viewed by the scientific realist not as necessary truths but as part of the only scientific explanation of the success of science, and hence as part of any adequate scientific description of science and its relations to its objects. (Putnam 1975a, p. 73)

Putnam sees the miracle argument as a scientific explanation, indeed the only scientific explanation, of the success of science. But in what sense is the explanation scientific? Is there a scientific law governing the development of mature science underlying some such explanation? Or is there some causal power which tends to make mature scientific theories true? This is very doubtful on both counts.

The superficial resemblance notwithstanding, Putnam's argument is very different from Lespieau's. Lespieau's argument can be elaborated along the lines explicitly developed by Perrin a decade or so earlier in favour of the existence of atoms on the grounds of the many independent methods for determining Avogadro's number (the number of atoms or molecules in a gram molecule of a substance) all giving more or less the same figure. And it can be made more specific, applying to molecules with a particular constitution and structure which explains so many different phenomena. This is not to say that there are not recalcitrant cases: many aspects of the microstructure of liquid water remain a mystery. But how do the successes of Darwin's theory of evolution, the law of mass action and the special theory of relativity bear on one another so that they might pull together to yield Putnam's conclusion? Darwin's and Wallace's theory doesn't provide independent support for the law of mass action and nor does the special theory of relativity, which rather calls into question the preservation of mass. A more reasonable view would seem to be that we should rest content with the accumulation of fallible knowledge and sensibly deploy whatever the fruits of progress may be.

The miracle argument is often presented as a general case of a form of scientific inference for which C. S. Peirce coined the term *abduction* and which is now also known as *Inference to the Best Explanation*. According to this scheme, alternative possible explanations of any phenomena are always available, even if sometimes too silly to need serious consideration, and we infer to the best of competing potential explanations. Sherlock Holmes infers that the butler did it because that makes best sense of all the available clues (Harman 1965). Until the first decades of the last century the universe was believed to be of constant size, with the stars occupying fixed positions. But the observed spectral lines from stars, compared with those derived from the spectra of elements observed on the Earth, were observed to be systematically shifted towards the longer wavelengths. Rather than postulate completely unknown elements in the stars, or other optical effects, the "red shift" was taken to be a Doppler effect caused by the source of light receding in an expanding universe. Again, Leverrier and Adams inferred the existence of Neptune because that gave the best account of the observed irregularities in Uranus' orbit (the alternative was to abandon Newton's laws). The same trick didn't work with the postulation of the planet Vulcan to explain the irregularities in Mercury's orbit, and Newton's theory of gravitation gave way to Einstein's general theory of relativity because that gave the best explanation.

Critics point out that inference to the best explanation is formally fallacious (committing a fallacy of deductive reasoning traditionally known as affirming the consequent). This may well be correct, but the same applies to any form of inductive inference (including the pessimistic induction!), which is deductively invalid.[6] The major flaw in this scheme is rather the presupposition that criteria of goodness of explanation are available a priori, antecedent to the actual justification and acceptance of scientific theories in terms of which scientific explanations are couched. The criteria determining one explanation to be better than another are often not settled in advance.

A case in point is provided by Galileo, who was astonished that Kepler assented to the "moon's dominion over the waters, to occult properties, and to such puerilities" (Galileo 1632, p. 462). He avoided what he took to be non-explanatory, occult properties in his own theory of how the earth's motion gives rise to the tides. But Galileo's appeal to tidal motions as a demonstration of the earth's motion was only as good as his own view was able to accommodate the evidence. Although many of his contemporaries agreed with him about Kepler's view, they thought Galileo's own theory no better. Incorporation of action at a distance was generally held to count against potential explanations, and Newton's explanation by such a notion wasn't accepted as plausible until after the overall balance of evidence making it the best candidate was appreciated. Disapproval of action at a distance was offset by the virtue of unity achieved by bringing together terrestrial and celestial phenomena within the scope of the same laws—a goal sought by Copernicus, Tycho Brahe and Galileo when they removed the Aristotelian division between sub- and superlunar regions of the universe. Greater precision counted decisively in favour of Newton's theory of gravitation. Newton considerably improved upon the tidal theory of Galileo, which allowed for only one daily tidal flow, thereby supporting the idea of the moon's dominion over the waters. Newton's theory also had the virtue of proposing what came, once the theory was accepted, to be seen as providing an explanatory causal mechanism in terms of a detailed account of the operation of forces. The formulation of criteria by which one explanation is judged better than another evolve alongside the development and acceptance of specific theories. Once the evidence had been taken to support Newton's theory over its competitors, the explanation of terrestrial and planetary motion could be grasped by mastering Newton's theory. What made it the best explanation was that it was the best-supported theory. Newton's theory of gravitation was radical insofar as it challenged previously accepted ideas of the merits of good explanation. But once accepted, it explained the evidence. The theory wasn't accepted because it provided the best explanation; it provided the best (correct) explanation because it was accepted.

The strength of abductive inference would depend in any particular case on the alternatives actually on offer. There may be nothing to say that any of them is really up to scratch and "the best among the historically given hypotheses ... may", as

---

[6]For a recent critique of inference to the best explanation from the perspective of inductive scepticism, see Weintraub (2017), who argues that it might well be "in a worse epistemic position than enumerative induction" (p. 188).

van Fraassen says, "well be the best of a bad lot" (1989, pp. 142–3).[7] Some of the examples given three paragraphs back to illustrate the idea of inference to the best explanation are misleading in this respect. They don't even involve the serious comparison of competing theories (the self-serving comment that the alternatives might be too silly to mention is tantamount to admitting as much). Abandoning Newton's laws in the face of the observed irregularities in Uranus' orbit was not a serious alternative so long as there was no alternative mechanical theory. This is clear from the fact that the same trick didn't work with the postulation of the planet Vulcan to explain the irregularities in Mercury's orbit in the course of the nineteenth century.[8] It wasn't before a serious alternative came along in the form of Einstein's general theory of relativity that this unsatisfactory state of affairs was resolved and the counter-evidence for Newton's theory could be deployed as evidence in favour of Einstein's. Then we had to relearn how to understand gravitation and how to explain the influence of the sun over the planets. Where serious alternatives are at issue, the candidate standing in closest agreement with the evidence is accepted on the basis of an eliminative induction. This is how Newton typically proceeded. He argued for his own theory of light as comprising a mixture of rays of differing refrangibility, for example, on the grounds that results of his own experiments showed the alternative theories of Grimaldi, Descartes and Hooke to be false and his own theory correct (Raftopoulos 1999). Holmes infers that the butler did it because by elimination, that is the only hypothesis consistent with all the evidence. From that we can conclude that the butler's actions explain the evidence.[9]

Some defenders of inference to the best explanation recognise its weakness. Schupbach has recently suggested that what he sees as the lack of "precise articulation and compelling defense" (2017, p. 39) of abductive inference can be addressed by construing it as inferring to hypotheses which render the phenomenon at issue less surprising (or more expected). Even if the phenomenon is not initially so surprising, in Schupbach's view, the inferred abductive hypothesis still makes it

---

[7] Van Fraassen goes on to make a case against scientific realism in general and the miracle argument in particular. He famously proposes an anti-realist view of science (discussed in Sect. 7.4) called constructive empiricism, the best argument for which is that "it makes better sense of science, and of scientific activity, than realism does and does so without inflationary metaphysics" (1980, p. 73). But the point in the text stands apart from the other things van Fraassen puts into this argument. If Mizrahi (2018) is correct, his argument can be turned on its head with his suggestion that it is an inference to the best explanation!

[8] Ad hoc modifications of Newton's inverse square law to accommodate the anomaly were tried, but these entailed problems with the perihelion motion of Earth and Venus and were not viewed as serious alternatives (Norton 2011, pp. 7–8).

[9] It is a trivial point that argument by elimination in this fashion is not deductively valid in the absence of a completeness premise to the effect that there are no further alternatives. And who, in Duhem's words, would "ever dare to assert that no other hypothesis is imaginable?" (1954, p. 190; cr. Sect. 6.4). But the argument is advanced, as Newton was acutely aware, as an inductive argument. The possibility of unconceived alternatives (*pace* Stanford 2018) doesn't detract from the support eliminative inductions like Newton's give for accepting the truth of their conclusions (cf. Chakravartty 2008).

less surprising. He goes on to relate this notion to probability in a tightening up of the proposal with an appropriate mathematical structure. The motivation may seem compelling in everyday cases like the plausible explanation of a crowded platform by the delayed arrival to a train, which we naturally infer. But such an ordinary notion hardly forms the basis of scientific inference. So far from rendering planetary and trajectile motion less surprising, for example, Newton's explanation in terms of gravitation was surprising and unexpected, not to say implausible, and only accepted as an explanation because of the evidence.

The explanation of how chemical elements combine to form molecules—an obvious and unsurprising fact once the existence of molecules was accepted—provides a good example of how explanation is sought in what has proved to be an empirically well-founded theory. The first workable theory of chemical affinity was provided in the latter half of the nineteenth century by the theory of thermodynamic potentials. Thermodynamics, a macroscopic theory still in use today, was originally conceived as a theory of the working of steam engines, whose relation to chemical affinity was far more surprising and unexpected than the well-known phenomenon of chemical affinity itself. An explanation in microscopic terms had to wait until well into the twentieth century, and is still subject to controversy. Something of the inadequate nineteenth-century attempts, illustrating the intractability of the problem within the classical mechanical framework, is described in Sect. 7.3.2. With the appearance of quantum theory in the late 1920s, the blatant paradoxes inherent in the attractive interaction of like-charged particles on Lewis's early theory of chemical bonding were disarmed and a framework became available for exploiting the changes in the structure of isolated atoms as they approach one another to form a chemical bond. On this basis, John Slater suggested in the 1930s that the cause of bonding is the electrostatic attraction between the positively charged nuclei and the negatively charged electron cloud built up by the accumulation of electron density between two nuclei as the atoms come together, creating stability compared with the isolated atoms. The idea was reinforced by Feynman (1939) and this understanding dominated the textbooks for many decades. But an alternative suggestion due to Hans Hellman, also dating back to the 1930s, was revived by Klaus Ruedenberg in the 1960s and has been coming into favour in more recent years. On this view, the stability arises because of the decrease in kinetic energy as electrons delocalise and can more easily move around a larger region. Here the competing explanations clearly could not have been formulated prior to the development of quantum mechanics, but are advanced within the framework of that theory. Moreover, champions of each of the two views do not argue for their preferred explanation by appeal to general criteria of explanatory goodness but in terms of the pros and cons of each as determined by the theory as they see it. Proponents of the electrostatic attraction view, for example, discounted kinetic energy pressure on the strength of the quantum-mechanical analogue of the classical virial theorem, whereas Rudenberg thinks this misguided and argues that it is necessary to apply variational analysis to the approach of initially separated atoms in the formation of a bond in order to isolate the causal factors (Needham 2014). This is not so much a dispute about which explanation is the better as which explanation is correct according to the theory. Adequate explanation is a matter of understanding the relevant application of theory that is supported by the body of accumulated evidence.[10]

Why would scientists be concerned to distinguish the correct explanation provided by theory from instrumentally adequate, and perhaps (as in this case) more easily grasped and therefore pedagogically more convenient, proposals, if not motivated by a concern for the truth? I take examples such as this of the quest for explanations properly grounded in theory supported by the evidence to illustrate a natural realist attitude actually found at work in the science community. This is a moderate realist attitude, not motivated by superscientific inference of the kind discussed in this section or the general semantic arguments discussed in the next. It is guided by an understanding of theories as critically consolidating past gains and not necessarily taken literally in their entirety as they are formulated, but implicitly restricted to particular domains of applicability and interpreted literally only with respect to their considered domain of applicability and up to the limits of experimentally justified precision. Further illustrations are developed in connection with the discussion in the next two sections in the coarse of steering between more extreme realist and antirealist views.

7.3   The Vicissitudes of Reference

Part of Putnam's miracle argument included the claim "that the same term can refer to the same thing even when it occurs in different theories". As an argument regarding scientific progress, this is particularly important and controversial where the theories at issue appeared at different stages over a relatively long period of time. Such diachronic comparisons can be quite difficult to assess, involving considerations pulling in different directions, and are not decided by perfunctory judgements based on trivial facts such as the use of the same term. Consider the term "atom", for example, which was introduced by the ancient Greeks. The fact that arguments of Aristotle's and of nineteenth century scientists against the intelligibility of atoms, which were influential in their time, do not work against atoms and elementary particles as understood since the early twentieth century speaks against preservation of reference. On the other hand, chemists have striven to maintain the insights of structural chemistry in conjunction with Daltonian ideas in the course of developing the theory of valency on the basis of quantum mechanics. Some aspects of the history of atomism since Dalton will be considered shortly.

---

[10]Another example of insisting on the correct explanation is to be found in the case Lubbe and Fonseca Guerra (2019, pp. 2766–7) and Lubbe et al. (2019) mount against the secondary electrostatic interaction model formulated in 1990 to explain differences in binding strengths between multiple hydrogen-bonded arrays. Although the model has been successful in predicting differences in experimentally determined binding strengths, theoreticians have sought to replace the arbitrarily oversimplified features of the model with a more accurate and sophisticated picture.

One idea Putnam was reacting against was an extreme form of holism (sometimes associated with Kuhn) according to which the meaning of a term is determined by the theoretical principles in which it features and from which it is concluded that any change in the theory entails a change in the meaning of the term. That is a non sequitur. Even if theories do determine the meaning of their terms, different theories may well determine the same meaning, just as a two independent sets of equations might determine the same values of one or more of the unknowns. This holistic thesis doesn't stand on its own ground[11] and Putnam's positive thesis is therefore not required for the purposes of rejecting this claim. Establishing the preservation of reference calls for an independent motivation and a much more detailed consideration of cases. Putnam illustrates his general idea with a thesis of the preservation of the extension of the "water" predicate since antiquity (typifying chemical substances and more generally, so-called natural kinds over the periods during which the corresponding terms have been in use). This is discussed in the next subsection. Then we turn to consider another case covering a far more restricted period: whether the use of the term "atom" by Dalton at the beginning of the nineteenth century has been preserved in the use of the term over a century later when quantum mechanics had been established by 1930. Finally, we take up the suggestion that there are some remnants of reference to be saved from terms like "caloric" and "ether" despite having been abandoned apparently by reason of lacking any reference.

7.3.1 Has "Water" Preserved Its Extension?

Putnam has proposed an externalist theory of reference, which has been influential in promoting the understanding of realism as successful reference. According to this theory, factors determining a predicate's extension (the set of things that it applies to) are independent of the thoughts of whoever might use the term at a particular time. As he succinctly puts it, " 'meanings' just ain't in the head" (Putnam 1975b, p. 227). The standard example is "water", which is supposed to hold of whatever it applies to even for those who know nothing of modern chemistry, be they our contemporaries or figures from history reaching all the way back to Aristotle and before. What determines whether something counts as water, and so falls under the "water" predicate, is whether it

> bears a certain sameness relation (say, x is the same liquid as y, or x is the same$_L$ as y) to most of the stuff I and other speakers in my linguistic community have on other occasions called 'water'. ... The key point is that the relation same$_L$ is a theoretical relation (Putnam 1975b, p. 225)

---

[11]Compare the way Feyerabend's "meaning variance" thesis fails in the face of Duhem's argument about the role of approximate truth in science (Sect. 6.6).

Putnam considers a thought experiment in which a substance with chemical constitution XYZ on a distant planet Twin Earth looks like water does to us. Superficial appearances might tempt us to say that such water-like quantities of material are water. But despite what Putnam takes to be the possibility of superficial indistinguishability, the matter is decided by the criterion of sameness, which Putnam says is being H_2O. The compositional formula of water was settled in 1850 by Williamson (who determined the common structure of alcohols and ethers by showing them to be of the water type, i.e., derived by substitution of hydrogen in water by alkyl groups—see footnote below on isomers). Although dating from the mid nineteenth century, being H_2O is what, according to Putnam, determines the import of the same$_L$ relation whenever "water" is used. Water is necessarily H_2O, whether the user of "water" knows it or not, "as if later theories in a mature science were, in general, better descriptions of the same entities that earlier theories referred to" (Putnam 1975b, p. 237). On this view, "water" or its Greek equivalent refers to the same stuff in Aristotle's usage as in ours, namely what stands to a paradigm sample in a sameness relation that neither he nor we (unless we know enough science) are able to explicate.

There is an issue about what Putnam and his followers mean by "is H_2O" and whether that accords with modern science. How far it is necessary to delve into modern chemical theory in order to give an adequate microscopic characterisation of a same substance relation applicable to water is a moot point. But understanding "H_2O" as simply giving the combining proportions of hydrogen and oxygen in the form of a compositional formula suffices to distinguish water from all other substances since it has no isomers.[12]

The distinction between substances and phases has greater bearing on Putnam's preservation thesis. One and the same chemical substance is typically to be found in the solid, liquid or gas state under different circumstances, sometimes exhibiting several phases (as chemists call these states) simultaneously. (Several solid phases of water are distinguished under conditions of high pressure.) The transformation of a quantity of liquid water into vapour by boiling doesn't convert it into a different substance and freezing converts it into the solid phase commonly called ice whilst remaining the same substance. Accordingly, the sense of "water" is independent of phase, applying to a particular quantity of a chemical substance irrespective of the phase(s) it happens to exhibit. This is essential for the proper understanding of the compositional claim that water is H_2O and must be entailed by any adequate interpretation of being H_2O.

---

[12]Isomers are distinct compounds with the same composition. Ethyl alcohol and dimethyl ether, for example, both have the compositional formula C_2H_6O, but are distinct substances with quite different physical and chemical properties. Their distinct modes of chemical reaction are marked by ascribing ethyl alcohol the structural formula C_2H_5OH and dimethyl ether the structural formula $(CH_3)_2O$. Williamson arrived at these structures in 1850 by showing ethyl alcohol to be derived from water by substitution of one equivalent of hydrogen by one equivalent of the ethyl radical, and dimethyl ether by substitution of each of the two equivalents of hydrogen in water by an equivalent of the methyl radical (Duhem 1902, Ch. 5).

Further, over a broad range of conditions, including those of normal temperature and pressure, water exhibits different phases from those of hydrogen and oxygen. These elements have boiling points that are much lower than water's, and are gases over a large part of the temperature range when water is liquid or solid (all comparisons taken at the same pressure). There is no gas in liquid or solid water and it doesn't make sense to say that water comprises hydrogen gas and oxygen gas. When claiming that water is composed of hydrogen and oxygen, all the substance terms are understood in the phase-independent sense. The relation of being the same substance as understood in modern chemistry is therefore not relativised to any particular phase; in particular, the relation of being the same substance water is not a matter of being the same liquid. Ice and steam are also water (H_2O) and H_2O (water) is not composed of hydrogen gas and oxygen gas.

When speaking of theory thoughts nowadays often go to the realm of microstructure. But the distinction between the macroscopic concepts of substance and phase is every bit a part of chemistry's theoretical heritage. It is central to the chemical thermodynamics developed by Gibbs in the 1870s that played a major role in opening up the new branch of physical chemistry towards the end of the nineteenth century. The discovery of new substances, either by separation from naturally occurring materials or by synthesis, that is characteristic of chemistry had begun in earnest earlier in the century, when chemists had learned to look for sharp, well-defined melting and boiling points as criteria of purity. These rules of thumb were incorporated into a broader theoretical picture as aspects of the phase rule—an important theorem of thermodynamics—relating the number of substances in a quantity of matter to the number of phases it exhibits (for further discussion, see Needham 2010). It soon made its mark by leading Roozeboom and van der Waals to the discovery of a new substance (Daub 1976), and similar discoveries followed in its wake. It should come as no surprise, then, that the emergence of the distinction between substance and phase and an appreciation of their connection was an important step in the development of the concept of chemical substance.

Since Putnam takes the claim that water is H_2O to be a relation between an ordinary, everyday notion and a scientifically developed concept, he focuses on an understanding of water that has some basis in everyday usage. This notion is, in fact, a reflection an older one featuring in chemical theory, which may have been connected with ordinary usage in the past but has certainly left traces in modern everyday usage. Substances haven't always been understood as they are now. Aristotle's conception of substances builds on a somewhat different view of elements and compounds from the modern one. It exerted an influence on how such concepts were understood right up to the time of Lavoisier, even if he did reject some aspects of the Aristotelian view. Aristotle had no truck with atomism and Lavoisier, for reasons of his own (see Sect. 7.3.2 below), followed him on this point. Matter was taken to be continuous and substance properties were taken to apply to all of what they apply to: "any part of . . . a compound is the same as the whole, just as any part of water is water" (*De Generatione et Corruptione*, 328a10f.; in Aristotle 1984). The defining properties of elements endowed them with the capacity to react with one another, being so modified in the process that they were not retained by any parts of the resulting compounds. These elemental defining properties were taken to endow the respective elements with the property of being solid (in the case of the Aristotelian element earth), being liquid (in the case of the Aristotelian element water) and being gas (in the case of the Aristotelian element air). So water as Aristotle understood it differs from what modern chemistry understands as water not only in being one of the elements (from which all other substances are derived by combination) but also in being liquid and not, as we would say, independent of phase. Aristotle understood the evaporation of water as a transmutation into another substance, air. (For more details, see Needham 2009a).

The general conception of substances as phase-bound persisted into the Enlightenment. The influence of this conception is evident in Joseph Black's notion of latent heat, which he introduced in 1761. He evidently thought that supplying latent heat involved chemical reactions transforming ice into another substance, water, and water into yet another substance, vapour. Lavoisier coined the expression "caloric" for the heat substance, in terms of which these reactions might be represented by

$$ice + caloric \rightarrow water,$$
$$water + caloric \rightarrow steam.$$

After each stage of combination of caloric to form a new substance, further heating transfers free caloric, uncombined with any other substance, which leads to the increase in a body's degree of warmth and a gas's expansion as it accumulates. Black was clearly still in the grip of the Aristotelian conception of substance as necessarily connected with a certain phase: to change the phase is to change the substance.

Lavoisier adopted this idea in his own conception of substances, despite his reprimands concerning the Aristotelian elements (Lavoisier 1789, p. xxiii). He apparently retained one of the Aristotelian elements, listing caloric as the "element of heat or fire" (Lavoisier 1789, p. 175). This element "becomes fixed in bodies ... [and] acts upon them with a repulsive force, from which, or from its accumulation in bodies to a greater or lesser degree, the transformation of solids into fluids, and of fluids to aeriform elasticity, is entirely owing" (1789, p. 185). He goes on to define gas as "this aeriform state of bodies produced by a sufficient accumulation of caloric". Air, on the other hand, is analysed into components. During the calcination of mercury, "air is decomposed, and the base of its respirable part is fixed and combined with the mercury ... But ... [a]s the calcination lasts during several days, the disengagement of caloric and light ... [is] not ... perceptible" (1789, p. 38). The base of the respirable part is called oxygen, that of the remainder azote or nitrogen (1789, pp. 51–3). Thus, Lavoisier's element oxygen is the base of oxygen gas, and is what combines with caloric to form the compound which is the gas.[13]

---

[13] As an aside from the "same extension" argument, but linking up with the earlier discussion of Lavoisier (Sect. 3.2), Kuhn is not quite right to say that the conception of substances "changed only after Lavoisier's new chemistry had persuaded scientists that gases were simply a particular

Similarly, the new element base of hydrogen (1789, p. 198) is not to be confused with the compound hydrogen gas, listed under binary compounds formed with base of hydrogen (Needham 2012, pp. 276–9).

Far from having the same extension, then, the present-day concept of water applies to much that wouldn't have counted as water for Aristotle, namely all the H_2O in the solid and gas phase. And on Putnam's understanding of the claim "Water is H_2O" as relating the term "water" on the everyday understanding of a substance in the liquid phase to the term "H_2O" as understood in contemporary science, the claim is incorrect: there is no equivalence or identity between the terms. Aristotle may well have counted as water substances in the liquid phase that would now be considered distinct from water. But the additions to the extension are sufficient to contradict the preservation claim about the extension of "water". There was a substantial change of meaning in the concept of a chemical substance from the phase-dependent to the phase-independent notion heralded by the recognition, with the law of definite proportions, of the fixed elemental composition of compounds. The concept received its final form after the replacement of the concept of heating as the transference of caloric by the thermodynamic conception of a process involving change of energy in Gibbs' chemical thermodynamics in the late 1870s.

There is no suggestion of incommensurability here. Broadly understood, what Aristotle counted as water was a certain kind of substance in the liquid phase (i.e. a quantity of matter which is all water and all liquid), and this is determined by modern criteria of sameness of substance and sameness of phase. This substance concept as used by our forefathers may be defined in terms of concepts now in use, but preservation of reference (fixity of the extension of a predicate expressing the concept) doesn't follow from this; on the contrary.

There is a question about how mixtures are understood and how that bears on what belongs to the extension of a substance term. Is there any water in the sea? Aristotle knew that salt could be obtained from sea water but not from fresh water, and he took homogeneity to be a criterion of comprising a single substance. He would therefore have denied, presumably, that there actually is any water in a quantity of matter when it is sea water, whereas the modern idiom speaks of it as present in a solution (but see Earley 2005). This problem is exacerbated by the fact that ideas and standards of purity have undergone a radical development over this period.

Finally, Putnam's speaking of the preservation of a predicate's extension over millennia implicitly assumes that there is a single form of predicate appropriate for the regimentation or formalisation in logical terms of what is expressed with the term "water" and other substance terms. It takes no account of different forms

---

physical state of substances which could also exist in the solid and liquid form" (1978, p. 18). Although undoubtedly a *stage* on the way to the modern, phase-independent sense of substance, these quotations clearly show that Lavoisier treated liquid and gaseous oxygen as compounds of, and so different substances from, the substance base of oxygen, and so on for all his elements. There was no sudden "paradigm shift", but a gradual change of view based of new empirical studies.

appropriate for different eras. Views have in fact differed on the issue of whether various substances are understood to undergo transformation into other substances. On Lavoisier's understanding of elements as the final products of analysis, they were indestructible. This view held sway until the discovery of radioactivity, in which some elements spontaneously decompose into other elements with the emission of α-, β- or γ-rays. The idea was generally abandoned with the understanding that elements are generated and destroyed in stars. Thus, whereas his notion of oxygen calls for a unary predicate, ours calls for a dyadic predicate applying to a quantity of matter and a time, allowing for matter that is oxygen at one time not to be at another. The idea that compounds comprise immutable elements, according to some historians, predates Lavoisier, going back at least to Geoffroy at the beginning of the eighteenth century (Klein 1994). From this time up to Cavendish's synthesis of water from hydrogen and oxygen in 1784, showing it is not an element, a unary predicate might have been appropriate. But Aristotle, who took water to be an element, thought the elements transformable into one another and into the matter of non-elemental substances as a result of mixing. On both Aristotle's and Lavoisier's views, then, but for difference reasons, the term "water" would be appropriately captured by a dyadic predicate allowing for the possibility that what is a particular substance kind at one time is not at another. The Stoics, on the other hand, maintained in opposition to Aristotle that elements are preserved in a mixt (a homogeneous mixture—the ancients didn't make the distinction between compounds and solutions that became firmly established in the eighteenth century), and were not transformed as Aristotle would have it. But they maintained, with Aristotle, that water was an element, and so their conception would be appropriately expressed by a unary predicate. Apart from difficulties discussed above, then, claiming that the "reference" of "water" is an extension that has remained unchanged since time immemorial runs roughshod over significant differences of view that a more sensitive treatment would reflect (Needham 2018, pp. 350–1).

7.3.2 Has "Atom" Preserved Its Extension from Dalton's Time?

Democritus suggested that the changing features of matter observed at the macroscopic level could be explained in terms of motions of intrinsically unchanging atoms, which were too small to be perceived individually but when brought together in variously arranged agglomerations constituted macroscopic bodies. He offered no details of how particular macroscopic changes or stable features were related to the swerving atoms, and although thinkers returned to such speculations in the course of the ensuing millennia, they remained just that: speculations of a very high degree of generality. Modern school curricula make no mention of them. But pupils do learn about Daltonian atomism in connection with the laws of constant and multiple proportions on which the modern conception of chemical substance is based. Is it reasonable to say that the modern conception of the atom goes back, not to antiquity, but to Dalton in the first decades of the nineteenth century? The concept of the atom has become very sophisticated; but is it fair to say that it is essentially the same notion that has been developed and refined in the intervening period?

A crucial issue here is that of the empirical justification of the atomic theory. The speculative theories from antiquity up to the enlightenment are now generally held to lack empirical grounding and to have played no role in whatever empirically justified advances that were made in the understanding of the nature of matter up to the nineteenth century. Towards the end of the eighteenth century Lavoisier, for example, claimed

> ... if, by the term elements, we mean to express those simple and indivisible atoms of which matter is composed, it is extremely probable we know nothing at all about them; but if we apply the term elements, or principles of bodies, to express our idea of the last point which analysis is capable of reaching, we must admit, as elements, all the substances into which we are capable, by any means, to reduce bodies by decomposition. (Lavoisier 1789 [1965], p. xxiv)

The modern theory of the atom is empirically justified; indeed, many aspects of it are so strange that it would be difficult to see why they should be accepted other than by sheer force of the empirical evidence. Surely, a necessary condition for the preservation of reference of the term "atom" from Dalton's day to ours is that the conceptions at issue have enjoyed empirical support, and perhaps even that this relation of empirical support has been preserved. The question is, then, whether Dalton's atomism was empirically justified. It is by no means clear that the answer is affirmative.

Dalton's atoms were indivisible extended balls each with a weight characteristic of the element of which they were the smallest possible portions. Boyle had previously proposed a corpuscular account of matter with a view to distinguishing the minimal units of the elements from one another in terms of an internal structure built up from more elementary corpuscles. This allowed for the transmutation of one element into another, which had been taken to be possible since Aristotle, but was ruled out by Lavoisier's notion of an element as the last product of analysis which Dalton carried over into his conception of atoms. Each Daltonian atom had its own measure of caloric, sometimes described as an atmosphere, sometimes taking the form of sticks normal to the spherical surface, in terms of which interatomic interactions were explained. Atoms of the same elemental kind repelled one another (the stick arrangements somehow facilitating this whilst allowing the caloric surroundings of unlike atoms to interpenetrate). This selective repulsion was to explain Dalton's law of partial pressures, according to which each distinct substance composing a quantity of gas exerts its own partial pressure as though the other substances were not present in the gas mixture, so that the total pressure exerted by the gas is the sum of the partial pressures.

Dalton claimed that his atomic hypothesis explained the law of constant proportions, which was finally established by Proust around this time against the last pocket of resistance in the form of Berthollet's objections. Berthollet compared compounds with what came to be called saturated solutions, whose composition is fixed by the circumstance of the solvent not being able to take up more of the solute. We think of zinc forming a compound by reaction with sulphuric acid, for example. But adding zinc to sulphuric acid is like adding sugar to water. At first, the sugar, like the zinc, disappears. But beyond a certain point, no more zinc is taken up into the liquid and it remains as a solid, just as continually adding more sugar to the water eventually leaves solid sugar once the power of the liquid to take up the solid into solution has been saturated. By Berthollet's lights, the combination proceeds, with the production of a homogeneous product, in both cases until the affinities are exhausted. Proust showed that whereas the composition of a saturated solution varies with temperature and pressure, compounds could be distinguished from solutions by virtue of their elemental composition remaining fixed whatever the ambient temperature and pressure, and whatever the chemical route by which they are produced.

It was a controversial question at the time whether the air, then known to contain oxygen and nitrogen, was a compound or a solution. This was settled by the fact that gases mix in all proportions, and unlike a compound, there is no fixed proportion of oxygen and nitrogen in a homogeneous sample of air. Dalton thought that the dissolution of one substance in another is "purely a mechanical effect", occasioning "no need to bring in chemical affinity" (Partington 1962, p. 773). What he thought of as the purely mechanical mixing of the constituents of air could be explained, as we have seen, on the basis of an atomic theory in which like atoms *repel* one another and the different substances exert their own partial pressures.

It could be said that the bare fact of constant proportions implies nothing about whether the division of matter would eventually come to an end with atoms or proceed indefinitely. It is consistent with either alternative and some further point would be needed to sway the balance (Needham 2004). Be this as it may, on the Daltonian hypothesis, combining proportions were supposed to be the ratio of the sums of the weights of all the atoms of each element in any given quantity of the compound, and thus reduce to the ratio of the sums of weights of atoms of different kinds in the minimal unit of the compound (the notion of a molecule, in which small numbers of atoms are bonded together, was yet to come). But combining proportions don't yield atomic weights (relative to hydrogen, say) without recourse to the number of atoms of each elemental kind in the minimal unit of the compound. For this purpose, Dalton introduced an assumption, the "rule of greatest simplicity", according to which one atom of each element is to be found in the unit. Lavoisier's determination of the composition of water as 85% oxygen and 15% hydrogen, for example, led Dalton to conclude that oxygen atoms are $5\frac{2}{3}$ the weight of hydrogen atoms, assuming in effect a binary formula, HO.

Dalton's law of multiple proportions[14] exacerbates the problem. Carbon and oxygen combine in different proportions to form carbonous oxide and carbonic-acid gas (what are now called carbon monoxide and carbon dioxide), for example, and there are several oxides of nitrogen. Dalton sought to accommodate this by assuming that one atom of one kind joins to one atom of another unless more than one compound is formed by the combination of two elements, in which case the

---

[14]When the same elements combine to form several different compounds, they do so in proportions which are simple, integral multiples of one another.

next simple combination arises from the union of one atom of the one kind with two atoms of the other. Dalton's contemporaries raised the obvious objection of arbitrariness. Realising something must be said about the nature of these atoms which would explain their postulated behaviour, Dalton responded with a suggestion about how atoms pack together. Assuming all atoms are spheres of the same size, he calculated that a maximum of 12 such spheres can pack around a given sphere. Importing the idea he postulated in connection with his law of partial pressures, that atoms of the same kind repel one another, the packing would be more stable the fewer atoms of a given kind pack around a given atom of another kind. A constitution of water corresponding to the binary formula HO (equal numbers of hydrogen and oxygen atoms) would minimise the internal repulsion and therefore be the most stable.

Dalton's atomic explanation of the law of constant proportions is therefore based on exactly the same principle that he used to explain the stability of air, i.e. of an ideal homogeneous solution. So Dalton simply did not rise to the challenge of explaining the difference between compounds and solutions, which is the nub of the Berthollet-Proust controversy. It was already apparent to Dalton's contemporary critics, Berzelius in particular, that Dalton offered no account of chemical *affinity*. But it is the concept of affinity that is the enduring notion that chemists of all ages have grappled with in trying to understand the state of combination of elements in compounds. Can we identify any contribution of Dalton's in the contemporary understanding of this issue?

There is much here that will not be recognised from the account of Dalton's atomism given in modern chemistry textbooks for schools. Daltonian atomism as usually presented nowadays is an abstraction from Dalton's atomism, shorn of the features he intended to explain how atoms act but were soon relinquished by science. Later on in the nineteenth century, with the development of the description of organic compounds with chemical formulas which were fixed without the need for Dalton's rule of simplicity (Duhem 1902; Needham 2008b), a common complaint against the atomic interpretation was that no independent features of atoms were ever presented which as much as hinted at how they might combine to form larger stable structures such as molecules. Features such as the valency of elements, well defined on the basis of the account of chemical formulas in macroscopic terms, were simply read into the atoms. The same goes for the extension of the essentially topological ordering of elements in compounds to three-dimensional structures which could distinguish between mirror images, posited towards the end of the nineteenth century to represent the distinction between optical isomers (pairs of compounds agreeing in all their chemical and physical properties with the exception of the direction in which their solutions rotated the plane of plane polarised light.) Those who took the representations to be of molecules made up of atoms had nothing to offer by way of suggesting how such structures could interact with polarised light to induce such rotations—even on a simple, qualitative basis.

Stanford has drawn on Dalton's atomism in support of his antirealist stance. He is right to say that the "profound differences between the central theoretical posit of Dalton's theory and *any* entity recognized by contemporary atomic theorizing might well lead us to deny that Dalton's atomic theory is even 'approximately' true" (2018, p. 217). But he is wrong to suggest that Dalton "achieved considerable empirical and practical successes" (loc. cit) with his atomic theory and fails to mention that Dalton's atomism, like the caloric theory, was controversial and by no means generally accepted by his contemporaries. For example, when awarding Dalton the Royal Society's medal in 1826, Humphrey Davy pointedly omitted Dalton's atomic theory when enumerating what he considered of value in Dalton's contributions. After mentioning several names involved in the history of the law of constant proportions, he says

> But let the merit of discovery be bestowed wherever it is due; and Mr. Dalton will be still pre-eminent in the history of the theory of definite proportions. He first laid down, clearly and numerically, the doctrine of multiples; and endeavoured to express, by simple numbers, the weights of the bodies believed to be elementary. His first views, from their boldness and peculiarity, met with but little attention; but they were discussed and supported by Drs. Thomson and Wollaston; and the table of chemical equivalents of this last gentleman, separates the practical part of the doctrine from the atomic or hypothetical part, and is worthy of the profound views and philosophical acumen and accuracy of the celebrated author. (Davy 1840, pp. 96–7)

There is therefore no warrant for Stanford's version of the pessimistic induction from Dalton's theory. Likewise, incommensurability is not an issue and there is no support for Kuhn's idea of scientific revolutions to be gained from this example. The idiosyncratic aspects of Dalton's atoms are straightforwardly described from a contemporary viewpoint and raise no problems of intelligibility for the modern reader.

Firm support for "the atomic hypothesis"—i.e. the thesis that matter is only apparently continuous on the macroscopic scale and has a very different structure at the microscopic level—was beginning to mount by the first years of the twentieth century.[15] Bohr's account of the atom in 1913 gave a promising explanation of electronic spectra. But it offered little that could explain how chemical bonding worked or provide a basis for the directional features of the putative bonds required by the atomic interpretation of structural formulas. This prompted G. N. Lewis (1916, 1917) to challenge Bohr's model on the grounds that, "assuming that the electron plays some kind of essential role in the linking together of the atoms within the molecule, and . . . no one conversant with the main facts of chemistry would deny the validity of this . . . [i]t seems inconceivable that electrons which have any part in determining the structure of such a molecule could possess proper motion" (1917, pp. 297–8). He thought electrons had fixed locations in the atom, "which may, in the case of chemical combination, become the joint property of the two atoms" (1917, p. 302), because that was the only way to account for the evidence from chemistry (and could also accommodate some evidence from physics).

---

[15]There was Perrin's confirmation of Einstein's theory of Brownian motion; Planck's theory of black-body radiation and its relation to the newly discovered electron; aspects of stereochemistry such as steric hindrance in reaction kinetics and stereo-specific reactions in sugar chemistry; and so on.

The representation of covalent bonds as static electrons was easy prey for Bohr (1965, p. 35) in his 1922 Nobel acceptance speech, but he had no account of his own to offer. Lewis came to see the possibility of a reconciliation when Bohr abandoned the idea of electrons revolving in rings in favour of electrons in shells, but remained clear about the status of the chemical evidence:

> No generalization of science, even if we include those capable of exact mathematical statement, has ever achieved a greater success in assembling in simple form a multitude of heterogeneous observations than this group of ideas which we call structural theory. (Lewis 1923 [1966], pp. 20–1)

This called for a notion of a bond which could certainly be more deeply understood, but whose status he took to be established:

> The valence of an atom in an organic molecule represents the fixed number of bonds which tie this atom to other atoms. Moreover in the mind of the organic chemist the chemical bond is no mere abstraction; it is a definite physical reality, a something which binds atom to atom. Although the nature of the tie remained mysterious, yet the hypothesis of the bond was amply justified by the signal adequacy of the simple theory of molecular structure to which it gave rise. (Lewis 1923 [1966], p. 67)

Fruitful theories of chemical reaction mechanisms were soon developed by Langmuir, Robinson and others in terms of ideas about electrons in atoms in molecules that Lewis suggested, even if they couldn't at first be related to radioactivity, spectroscopy and scattering experiments that provided the first evidence of atoms. But applications of the new quantum mechanics from 1927 to some simple tractable cases pointed more firmly to links with physics, from which two schools of quantum chemists emerged. Insights from Heitler and London's (1927) solution to the diatomic hydrogen molecule were developed in the Valence Bond (VB) method by Linus Pauling. Here the approximation procedures necessary for the application of quantum mechanics were modulated in order to construe molecules as built up from individual atoms. This facilitated the interpretation of quantum mechanical results in terms of ideas from the classical structural theory of organic chemistry with which chemists were familiar and Lewis had sought to retain. This bridge between classical and quantum chemistry initially led the majority of chemists to favour the VB method over the quantum mechanically more purist Molecular Orbital (MO) approach developed by Robert Mulliken and Friedrich Hund. The MO theory threatened the classical idea of a bond as a localised material connecting link between specific pairs of atoms in a molecule that Lewis had singled out and Pauling had successfully retained in the VB approach. Molecular orbitals were delocalised, defined over the whole molecule in which the individual atoms were no longer seen as the integral parts that they were on the classical theory. As the MO approach gained the upper hand after the 1960s, largely because of the more tractable calculations that were becoming feasible when formulated in MO terms for ever more complex systems,[16] it seemed that the world of quantum chemistry was becoming ever more foreign to the ideas which chemists had found so useful.

---

[16]For a more general comparison, see Kutzelnigg (1996, pp. 576–8).

Quantum chemists developing Mulliken's molecular orbital approach have become increasingly sceptical about the value of sticking to these old ideas. C. A. Coulson, for example, went so far as to declare "a chemical bond is not a real thing: it does not exist: no-one has ever seen it, no-one ever can. It is a figment of our own imagination" (Coulson 1955, p. 2084). Mulliken had had his doubts from the outset:

> In the molecular point of view advanced here, the existence of the molecule as a distinct individual built up of nuclei and electrons is emphasized, whereas according to the usual atomic point of view the molecule is regarded as composed of atoms or of ions held together by valence bonds. From the molecular point of view, it is a matter of secondary importance to determine through what intermediate mechanism (union of atoms or ions) the finished molecule is most conveniently reached. It is really not necessary to think of valence bonds as existing in the molecule. (Mulliken 1931, p. 369)

These confrontations notwithstanding, later developments clearly suggest that the situation is one of Duhemian progress rather than Kuhnian revolution. Something of a rapprochement between classical chemistry and modern quantum chemistry has emerged in more recent times. Classical bonds correspond to what quantum chemists call a localised description of the molecular orbitals occupied by electrons. Although theory always provides a delocalised description, equivalent localised descriptions are sometimes available, namely when Hund's criteria[17] are satisfied (Kutzelnigg 1996, pp. 578–9). However, there are cases where the criteria are not satisfied and no such localised description is available, corresponding to delocalised bonding that the classical view failed to recognise, such as is found in compounds like benzene and MgH_2, as well as ions such as CO_3^{2+} and BeH_2^+.

As this makes abundantly clear, there is a great deal counting against any suggestion of preservation by modern chemistry of the atoms of Dalton himself or those of the abstraction of Daltonian atomism which were incorporated into the classical structural theory of organic chemistry. It is not just a matter of allowing for divisibility into electrons and constituents of the nucleus, which is radical enough a change. Even Lewis's idea that the evidence ensures that bonds must survive any changes in the detailed understanding of chemical combination has come under threat. On the other hand, appeal to a general idea of discontinuity at the microlevel is too vague and misleading to support the preservation thesis. Quantum mechanics is a continuous theory as far as space and time are concerned; there are no abrupt boundaries enclosing elementary particles and separating them from the rest of the universe. It is energy that changes in discontinuous steps, the consequences of which are felt at the microscale.

The atom in chemistry should be familiar to many readers and one of relatively few examples of a term that has been in longstanding use in science which has seen a considerable change over the last two centuries. The same cannot be said about the term "planet", for example, which has not undergone anything like the same kind

---

[17]When the number of bound neighbours, the number of available valence electrons and the number of valence atomic orbitals participating in the bonding are all equal. In BeH_2 these numbers are 2 for Be and 1 for each H atom.

of change of usage. The recent redefinition of the term excludes bodies like Pluto, which was counted as a planet when first discovered in 1930 but is now known to be much smaller than originally thought and unable to affect Neptune's orbit, and is not significantly different from many of the other bodies in the Kuiper belt. This leaves untouched the identification of the planets known to the Greeks—Mercury, Venus, Mars, Jupiter and Saturn—with the planets known by these names today. The same could be said about Earth, even if it has lost its status as the centre of the universe. There are serious doubts about whether anything for which the term "atom" is used today corresponds to anything that Dalton understood the term to apply to.

7.3.3 Are There Remnants of Reference of Apparently Abandoned Terms?

Premises of Laudan's pessimistic induction argument claimed that there are no such things as ether and caloric; many terms of bygone science like "ether" and "caloric" don't, in fact, refer to anything. Or so it seems. But some authors have suggested that some of the entities postulated by past science and now generally thought not to exist have in fact survived in other guises. Phlogiston, for example, was supposed not only to explain inflammability but also the general property of being metallic. This latter feature, for which Lavoisier's oxygen theory had no explanation, is an example of Kuhn loss. Today, metals are understood to have the general feature of being made up of positive metal ions and a body of electrons completely detached from these ions and effectively behaving like a gas. This electron gas occupies states very close in energy that explains the characteristic features of malleability, ductility and lustre of metals. It is tempting to liken the electron gas to phlogiston and the positive ions to the metallic calx that remained after heating the metal in air according to the phlogiston theory (Akeroyd 2003, p. 298). But the striking analogy (which is perhaps all Akeroyd points to) hardly amounts to preservation of reference. The electron gas doesn't explain inflammability even in metals, let alone the many other substances in which there is no electron gas, and positive metal ions are not what remains after heating metals in air but a metal oxide.

Stathis Psillos (1999, pp. 296–9) claims that the luminiferous ether in fact refers, and only the name has been abandoned. What determines reference, according to Psillos, is a "core causal description", which in the case of the ether involves a dynamical structure with two sets of properties:

> The first set … were, broadly speaking, kinematical. Given that it was experimentally known that light propagates with finite velocity, its laws of propagation should be medium-based as opposed to being based on an action-at-a-distance. The second set of properties was, broadly speaking, dynamical: the luminiferous ether was the repository of potential and kinetic energy during the light-propagation. (Psillos 1999, p. 296)

Psillos then argues that Maxwell's electromagnetic field satisfies the core causal description associated with the ether, so that "the kind-constitutive properties through and for which the ether was posited were 'carried over' to the conception of the electromagnetic field ... [and] the denotations of the terms 'ether' and 'field' were (broadly speaking) entities which shared some fundamental properties by virtue of which they played the causal role they were ascribed" (loc. cit.).

Two reasons speak against any such identification. First, the point of the ether theory was to provide a mechanical explanation of the propagation of light waves once it was established that light was a wave phenomenon rather than consisting of particles. Nineteenth-century scientists could not understand how waves could be propagated if not via a medium, like waves on water and sound waves in the air, although it was puzzling why the wave motion should be transverse rather than longitudinal as in the case of sound waves. Identifying the electromagnetic field with the ether simply disregards this fundamental motivation for the ether. Maxwell's equations came to be understood as describing the propagation of light in terms of a changing electric field, which generates a magnetic field that in turn generates an electric field as it waxes and wanes, and so on, without needing to appeal to a mechanical ether. Second, the famous null result of the Michelson-Morley experiment showed that the ether does not exist.[18] Surely the point of realism is that existential statements should be taken literally. This applies just as much to negative existential claims as to positive ones, whose significance would be impaired if negative existential claims were not also taken literally. So the realist shouldn't shrink from denying the existence of entities when the occasion calls for it, and doesn't need to save them at all costs. As we saw, Laudan's pessimistic induction poses no threat, so there is no need to try to undermine the premises of his argument in this way. If this shows that the notion of a core causal description is not adequate to establish the reference of a term, then so much the worse for that notion. Whatever the merits of the causal theory of names advanced in the philosophy of language as the basis of the reference of proper names, it doesn't seem to offer any clear guiding lines for the use of general terms in scientific theories.

Psillos has also attempted to save caloric on the same general grounds of reference to the causes of the phenomena they allegedly explained. Specifically, "the laws of the caloric theory can be deemed to be approximately true independently of the referential failure of 'caloric' " (Psillos 1999, p. 127). The fact that scientists think there is no caloric because heat is not conserved, undermining an essential feature of caloric theory, counts very strongly against this. The distinction between warmth (measured by temperature) and heating came to be understood in terms of alternative ontological presuppositions, as a distinction between a feature of the state of a body of matter or radiation and the property of a process. At the same time, the truth was recognised of Lavoisier's premonition (mentioned in Sect. 7.2) that, even if "[n]othing is more readily conceived, than that one body should become elastic by entering into combination with another body possessed of that quality",

---

[18]Ohanian (1988, pp. 158–64) reviews the increasingly accurate experimental evidence since the original experiment, with varying experimental procedures, which considerably lowers the upper limit to the speed of an ether wind that would go undetected—clearly a point of enduring significance.

> We must allow that this is only an explanation of elasticity, by an assumption of elasticity, and that we thus only remove the difficulty one step farther, and that the nature of elasticity, and the reason for caloric being elastic, remains still unexplained. (Lavoisier 1789, p. 22)

Late nineteenth-century proponents of a "purely mechanical" account of the workings of the ether were sensitive to the need to avoid this kind of circularity (Hunt 1991, pp. 100–5). But the failure to provide any satisfactory account of electric charge on this basis reinforced the acceptance of the electron and sank another nail into the coffin of the ether.

Chang (2003) offers a convincing rebuttal of Psillos's claim about caloric by pursuing other historical details, but I can't agree with him that "the real history of the caloric theory ... support[s] Laudan's pessimistic induction" (p. 910). Defeating extreme realism doesn't amount to an argument for Laudan's antirealism because these are not the only positions in the field. On the contrary, most scientists do believe in electrons, protons and neutrons, and it is part and parcel of this natural realist position to affirm negative existential claims such as there being no caloric and no ether. The negative existential claims don't support an antirealist view.

It should be clear from the earlier discussion that Dalton's atomism cannot be saved by Psillos's strategy of understanding terms to refer to the causes of particular phenomena. Even if it is thought that Dalton's atomism explains the law of constant proportions, he certainly didn't offer an explanation of the source of chemical combination distinguishing compounds from solutions (what he took to be mere mechanical mixtures).

A variation on the preservation of reference is the structural realism first advanced by Poincaré (see Giedymin 1982). Poincaré was impressed by the continuity of mathematical structures which survive amid the devastation of particular theories dealing with particular kinds of entity.

> No theory seemed established on firmer ground than Fresnel's, which attributed light to the movements of the ether. Then if Maxwell's theory is to-day preferred, does that mean that Fresnel's work was in vain? No; for Fresnel's object was not to know whether there really is an ether, if it is or is not formed of atoms, if these atoms really move in this way or that; his object was to predict optical phenomena. (Poincaré 1902 [1952], p. 160)

What is preserved, it seems, is the empirical facts, much as maintained on van Fraassen's (1980) antirealist view (see Sect. 7.4). But Poincaré intended something deeper, for he continues

> The differential equations are always true, they may be integrated by the same methods, and the results of this integration still preserve their value. It cannot be said that this is reducing physical theories to simple practical recipes; these equations express relations, and if the equations remain true, it is because the relations preserve their reality. They teach us now, as they did then, that there is such and such a relation between this thing and that; only, the something which we then called motion, we now call electric current. The true relations between these real objects are the only reality we can attain, and the sole condition is that the same relations shall exist between these objects as between the images we are forced to put in their place. (p. 161)

These relations, expressed in what Poincaré called principles, are abstract. They were discovered in the course of developing advanced mathematical techniques such as Hamilton's wave mechanics, which was an attempt during the first half of the nineteenth century to distinguish common features in particle and wave theories of light (Giedymin 1982, Ch. 2). Such common principles can take the form of a common content invariant under a set of transformations and express "indifferent hypotheses". They are what might possibly be saved from the ruins of a falsified theory when more specific postulates fall away—principles that are indifferent to one or another set of specific postulates.

Paul Dirac spoke in much the same vein:

> Quantum mechanics was built up on a foundation of analogy with the Hamiltonian theory of classical mechanics. This is because the classical notion of canonical coordinates and momenta was found to be one with a very simple quantum analogue, as a result of which the whole of the classical Hamiltonian theory, which is just a structure built up on this notation, could be taken over in all its details into quantum mechanics. (From a 1933 paper of Dirac's, quoted by Bukulich 2008, p. 52)

As early as 1925, Dirac maintained that "the correspondence between the quantum and classical theories lies not so much in the limiting agreement when $h \to 0$ as in the fact that the mathematical operators in the two theories obey in many cases the same laws" (quoted by Bukulich 2008, p. 62). Bukulich goes on to say that Dirac had a "fundamental belief in the unity of physics ... in which the same 'basic structures' reappear (suitably adapted) in all the different branches. A branch of physics formulated without equations of motion will, in his view, remain disconnected from the rest of physics" (p. 60). This is based on an intuition expressed in a 1970 paper that "A theory with mathematical beauty is more likely to be correct than an ugly one that fits some experimental data" (loc. cit.). Heisenberg ridiculed any such suggestion, as Bukulich elaborates. But Dirac staunchly maintained that "much of the beauty that quantum mechanics has is 'inherited' from classical mechanics ... [because] a large part of the formal structure of classical mechanics is preserved in the transition to quantum theory" (p. 61). Even if this is right, however, it doesn't imply that what quantum theory is about is nothing but the structure preserved from classical mechanics. It could well be glossed as saying that entities of the classical and the quantum worlds have much in common. Much more is required to justify the thesis that there are no such entities, as the structural realist maintains.

Structural realism has been more recently taken up by Worrall (1989). In a general review, Ladyman (2009) is careful to distinguish between the epistemological claim that nothing but structure can be known and the ontological claim that structure is all there is. Ontological structural realism has found a new source of motivation as a way of dealing with indistinguishable particles of quantum mechanics. There is no hint, however, as to how it might be applicable to chemistry or biology where theories don't have the kind of mathematical form characteristic of theoretical physics. Without a leap of faith to embrace reductionism, this considerably restricts its scope. Even in the primary area of application, the central contention, that formal structures are what prove to be the lasting features of scientific discoveries, has been criticised for want of a convincing argument that structure can stand alone rather than being the structure of something and a precisely defined notion of structure applicable in significant cases. Moreover, if it is not just a thesis about the preservation of observable, phenomenological laws, it could be countered that the mathematical similarities are imposed for the purpose of developing a theory along familiar lines. Circles were a central feature of ancient and medieval astronomical theories. But this is a reflection of the restricted methods available at the time for the construction of a description of a planetary orbit rather than an enduring feature of physical reality (Price 1959). What remains true is the approximate truth of the planetary orbits as then described within the limits of error then feasible. Even if some structural feature which turns out to be an enduring trait could be characterised, it would remain to show that it constitutes the whole theory and that it is essential in the sense that no other theory formulation lacking this feature could possibly be adequate to the accumulated evidence.

7.4 Underdetermination

Structural realism seeks to isolate a part of current theory that can be distinguished from the remainder and whose significance can be promoted in the service of the proponent's view, in this case a variety of realism. Underdetermination is a thesis that seeks to distinguish a part of current theory, on the strength of which the theory is said to be *empirically adequate*, in the service of a view opposed to realism known as *constructive empiricism*. Needless to say, this is a different part, distinguished in terms of what is observable. It should be said, however, that the relevant sense of "observable" owes more to the strictly philosophical tradition of empiricism than to how the term is used in modern science (see Bogen and Woodward 1988), although it may be more closely allied with scientists' usage in the historical origins of the doctrine.

The claim that where one theory is empirically adequate, there will be a radically different theory such that evidence couldn't possibly justify choosing between them is known as the underdetermination of theory by evidence. The idea goes back to an ancient and medieval tradition in astronomy which distinguished between true physical hypotheses explaining why celestial bodies move as they do, and mathematical hypotheses formulated for the sole purpose of predicting future movements. Osiander, a theologian who oversaw the publication of *De Revolutionibus* during Copernicus' final illness, famously appealed to this distinction in a preface (about which Copernicus presumably knew nothing) to argue that accepting the mathematical virtues of the way Copernicus' theory deals with observations didn't imply accepting the literal truth of the theory, which he thought was decided by other considerations. The Greeks were aware that the same orbital motions could be constructed in different ways from different systems of circles. It was apparent to Hipparchus (c.190–c.120BC), for example, that two explanations could be given of the fact that the Sun takes almost 6 days longer to traverse the ecliptic from vernal to autumn equinox than it does to return through the other 180°

during winter.[19] Conformation of the observed solar motion to the predicted orbit therefore provided no greater evidence for the one construction over the other; they were observationally equivalent hypotheses which were underdetermined by the evidence. According to the argument from the underdetermination of theory, there are always alternative, observationally equivalent theories to any given theory, and since we have no grounds for distinguishing between observationally equivalent theories, we have no grounds for believing in any one rather than any of the others. But we can't believe them all since they are mutually inconsistent and the antirealist says we should merely hold that a theory is empirically adequate, committing only to its observational consequences.

So much for the general idea. The example of the Sun's motion around the ecliptic might be taken merely to illustrate two essentially equivalent mathematical constructions of the same path rather than distinct mechanisms, each with its own mechanical devices, making conflicting claims about physical reality. More substantial examples are needed to bolster the claims of antirealism.

Capitalising on the fact that observations of planetary motions revealed only a relative motion between planet and observer (the Earth), Tycho Brahe (1546–1601) advocated a geocentric system—essentially that of Heracleides (388–315 BC)—observationally equivalent in respect of planetary motions to Copernicus' heliocentric system, in which the Sun revolves around the stationary Earth and the planetary orbits are centred on the Sun. In both systems, the planets move westwards seen against the background of the fixed stars (Kuhn 1957, p. 204). The Tychonic system thus retained all the mathematical advantages of the Copernican system without the drawbacks entailed by the moving Earth. The example illustrates a weakness in the appeal to observational equivalence to motivate constructive empiricism. The equivalence here holds only with respect to a limited class of observations concerning the kinematics of planetary motions. Tycho's argument was that the traditional arguments for a stationary Earth, based on terrestrial observations, decided the issue in favour of his theory. (This is why Galileo found it necessary to appeal to the tides, thinking his theory showed that the Earth really moves.) Examples of systems equivalent with respect to a limited range of observations may be relatively easy to give. But there is always a risk that the equivalence is not preserved under a broadening of the range of observations. The only way to avoid this and provide a motivation for antirealism based on underdetermination is to argue that there will always be alternatives to our total theory of the world which are observationally equivalent with respect to all possible observations. This is not established by giving simple examples like those we've seen so far, and it is difficult to see how an appropriate general argument could be given. Moreover, Quine—one of the authors to have followed through this line of

---

[19]Either the Sun can be placed on a small epicycle with a period of equal duration but opposite sense of rotation to that of the deferent, in which case a ratio of the radii of the smaller circle to the deferent of 0.03 accounts for the extra 6 days. Or the Sun can be directly ascribed a circular path whose centre is placed on an eccentric.

thought—points out (1975, pp. 320ff.) that there is in any case the distinct possibility that two such observationally equivalent tight theories ("tight" means free from gratuitous additions) could be shown to be merely trivial variants of one another by a systematic interchange of vocabulary. What are called electrons and protons in one theory, for example, are renamed protons and electrons, respectively, in the other; or what is a curve of a certain kind in one theory is a straight line in another.[20] Given two candidate total, observationally equivalent theories, it would have to be shown that there is no transformation of the one into the other that would show them to be essentially trivial variants, which may well be a nontrivial task.

Quine first introduced the term "underdetermination" in the course of the discussion following his "Two Dogmas" paper for a very radical thesis: "Any statement can be held true come what may, if we make drastic enough adjustments elsewhere in the system" (1951, p. 43). He subsequently dropped this radical claim, at least as a substantial thesis (Quine 1976). Underdetermination as discussed in Quine (1975) and later involves the less radical claim that two total theories of the world might be empirically equivalent and yet conflict in some of their claims. The idea that any sentence can form part of one empirically adequate theory and its negation part of an empirically equivalent theory is no longer an issue.

The more radical claim acquired the name "Duhem-Quine thesis" in the literature which has stuck (see, e.g., Papineau 1996, p. 7) despite Quine's change of heart and despite the fact that what Quine (1951, p. 41) attributed to Duhem is only the holism thesis discussed in the last chapter. Ariew (1984) clearly shows that there is no reason for thinking Quine should also have recognised that Duhem anticipated his radical underdetermination thesis, carefully distinguishing what is here called the radical underdetermination thesis—Ariew's "subthesis (ii)"—from anything Duhem claimed.

According to the underdetermination thesis Quine discussed in later writings, empirical equivalence requires consideration of *all possible* observations. Not only is the thesis to concern our total world theory and thus confront all actual evidence. The thesis is also to be distinguished from inductive uncertainty of theories confirmed only to a certain degree by evidence *actually available*. All parties are agreed that the nature of empirical research is such that even if a theory is supported by the evidence at one time, this is no guarantee that it will not be refuted later. Accordingly, the evidence available at any given time may be insufficient to decide between two seemingly distinct theories, and may continue indefinitely to be insufficient. If realism is to be criticised on the basis of the notion of underdetermination, this must be distinguished from inductive uncertainty.[21]

---

[20]This last example alludes to Poincaré's method of proving the relative consistency of Euclidean and non-Euclidean geometries by reinterpreting concepts in the one in terms of concepts in the other.

[21]In recent literature the term "underdermination" is sometimes used for the non-deductive, inductive relation between actual evidence and the hypothesis or theory it supports. We have seen this here with Stanford's claim that unconceived alternatives "underdetermine" hypotheses or theories at any given stage of the historical development of science (Sect. 7.2).

How is the notion of all possible observations to be explained? Quine (1960) introduced a notion of observation sentence, somewhat surprisingly after his rejection of what he called the positivist "reduction thesis" of empirical claims to sense data—what is supposedly immediately given to the senses and reported in observation sentences—in his Two Dogmas paper. But Quine's notion of an observation sentence was considerably weaker, being an unstructured sentence not presupposing a distinction between observation and theoretical predicates as did the positivists.[22] Incorporation into a theory required that observation sentences be analysed, resulting in sentences no longer treated holophrastically but assigned a logical structure in virtue of which they could enter into relations of logical consequence with theoretical principles and other sentences resulting from the analysis of observation sentences. Under such theoretical interpretation of observation sentences, times and places are introduced. All possible observations are obtained by considering all possible combinations of times and places replacing any actual such locations. Quine continued to modify and refine the idea under the force of criticisms, and the basis for the notion of observational equivalence remains, together with the problem of establishing a clear criterion distinguishing theories, among the principal problems of articulation of the underdetermination thesis.

A somewhat different tack is taken by Bas van Fraassen (1980), who argues for his constructive empiricism on the grounds of rampant underdetermination, undermining any justification in believing in the ontology of any particular theory as distinct from its empirically equivalent rivals. For him as for Quine, the notion of possible observation underlying empirical equivalence is not a matter of what is and is not theoretical. What he thinks makes a claim observable is that our best theories about our sensory capacities tell us that what is described, no matter how "theory laden" the description, can be observed. Observation concerns what is in principle observable by unaided human perception. What is observable may be a matter of degree. But van Fraassen thinks this point is met if the opposite ends of the spectrum, where the distinction between observable and unobservable are drawn, are distinct, despite the considerable area of vagueness in between and its being subject to ongoing scientific investigation. Crystal structure that gives rise to X-ray photographs is not observable because it can't be seen with the naked eye. Nor are atoms observable by electron microscopes. On the other hand, mountains on the other side of the moon are observable because an appropriately placed human would be able to see them, whereas there is no such possibility in the former cases. The change in entropy in the boiling of water and the change in potential energy as a stone falls to the ground presumably count as not noticeable by unaided perception on this view and therefore as not observable.[23]

---

[22] The idea was originally envisaged as the starting point of a linguist's project of radical translation of a foreign language for the first time, when the linguist offers sentences eliciting affirmation or denial when the native is confronted with what the linguist believes are corresponding circumstances.

[23] A change of entropy is measured by an amount of heating at a certain temperature and this heating could be detectable by the senses if confined within certain limits. Whether this means

On the strength of this delimitation of the observable, van Fraassen maintains that we are justified in believing theories to be *empirically adequate* exactly when they tally with what is observable. But since the observable substructure of any total theory cannot determine any one particular superstructure in which it is embeddable as distinct from any other, because it may be extended to such a superstructure in innumerable ways, the total theory is underdetermined. Accordingly, van Fraassen also maintains that belief in any one theory to the exclusion of its empirically adequate variants sharing the same observable substructure would be unjustified. He is not opposing the realist by advocating theoretical "atheism" and actually adopting an antirealist stance; he is merely a non-realist, agnostic on theory, offering an alternative to the realist's belief in the literal truth of scientifically justified theories. This is a purely abstract argument, which may point to no more than what the realist regards as pretty insignificant alternatives. Van Fraassen gives as his example of Newton's postulate of a universe in which the centre of the solar system is at absolute rest, which is contrasted with alternatives ascribing it some definite velocity, none of which the evidence points to uniquely. The realist responds that according to Newtonian mechanics, there is no absolute velocity.

I have yet to see even an attempt to illustrate the thesis in fields touching chemistry and biology. Once we leave the interesting examples from astronomy that provided the original motivation for the idea in search of a generally valid argument justifying the underdetermination thesis, it seems we leave the realms of real science in favour of fantastic and patently artificial constructions. Whatever solace they provide for the proponent of the thesis, the the realist remains unimpressed.

Van Fraassen's notion of observability has been the subject of much criticism. Ladyman (2000, 2004) questions whether it is consistent with van Fraassen's denial, in conformity with traditional empiricism, that any objective statements are made by modal and counterfactual claims. Where does this leave the distinction between the observable and the unobservable? Dinosaurs and centaurs are observable, but not actually observed. In the one case, this is perhaps the major reason why we don't think they have ever existed; in the other, we observe traces from which we infer the existence of prehistoric, frequently giant, reptiles which we would have seen were we sufficiently near at the right time. Empirical adequacy is a matter of tallying with all observ*able*, including all past and future as well as present, observations. We are therefore supposed, according to van Fraassen, to be able to take a justifiable stance on this without any commitment to the theory as a whole—without, that is, seeing the justification for such a stance as providing evidence for the theory.

Why the emphasis on naked eye vision? We have seen in Chap. 2 something of how science has progressed by extricating itself from the misleading results of vision by relying increasingly on instruments. Experimental data used in the confirmation of theory is not understood in science in terms of naked eye vision.

---

that the entropy change in a particular process, such as the separation of a gas mixture into pure components by diffusion through a semi-permeable membrane, or heating outside the limits of human detection, are observable is, to put it mildly, unclear.

Recent measurements obtained from the European space mission Planck of photons released after decoupling from atoms a few hundred thousand years after the Big Bang, for example, are said to be "the most ancient signal that has been *directly observed*." And "looking at binary stars, … the first *direct observation* of gravitational waves produced by the collision of two massive black holes has been reported." An important milestone in the development of elementary particle physics came in 1956 with the "[f]irst *direct observation* of a neutrino" (Iliopoulos 2017, pp. 20–1, p. 116; my emphasis). Realists have taken the kind of criticism levelled by Duhem against the idea of a crucial experiment to undermine a significant observation-theory distinction, and where any such distinction can at least be well-defined, to regard it as at best arbitrary. They do not accept that van Fraassen has succeeded in putting the burden of proof on them to justify belief in "unobservables". Where evidence is strong they take it to justify a theory. There is a place for caution. But its role is in assessing whether the evidence is entirely convincing or merely provides tentative support for the theory, and assessing the limits of experimental error; not in restricting belief to a claim of empirical adequacy.

7.5 Taking Stock

The advancement of scientific knowledge naturally encourages the adoption of a realist attitude in which empirically well-grounded theories are taken literally and the world is believed to be just as the theories say it is. Reflection leads us to temper this stance, not necessarily taking literally every facet of the theories as commonly formulated. But challenges based on general philosophical principles that argue for a much more selective attitude, accepting the practical or instrumental value of theories in facilitating technological innovation whilst viewing the deeper, systematic theoretical claims with scepticism have not carried conviction. The pessimistic induction projects into the future what proved to be scientific dead ends on the basis of an enumerative induction. But many of the examples on Laudan's list involve the postulation of some "underlying reality", like the ether for Maxwell's theory of electromagnetic radiation, which the equations don't really sanction. In any case, the putative premises of the induction were frequently only tentative proposals whose inadequacies were often appreciated and not the stalwart beliefs required for making the inference. Confidence was usually associated with collateral conditions circumscribing the possible error, making for more precise formulations of claims than is suggested by the sound bites that Laudan relies on. So it is misleading to say that Kepler's claims about the elliptical orbits of the planets were simply overthrown by Newton's claims based on his principle of gravitation. Kepler's claims were upheld within his explicitly formulated limits of error, and the elliptical orbits modified only within more tightly drawn margins of error—just as Newton's law of gravitation is still reliably used for many purposes and only replaced by relativistic trajectories within finer limits.

Scientific progress, amounting to the amassing of empirically well-supported claims, has always been accompanied by an aurora of speculation that sometimes bears fruit, sometimes not. Scientists at any particular time are not uniformly committed regarding what hindsight reveals to be mere speculation, but may be cautious or even sceptical. Focusing on those aspects that more prudent scientists are wary of doesn't give a fair picture the development of science.

Again, Kuhn's holistic ideas about the meaning of individual terms of a theory being uniquely dependent on the whole theory so that the slightest change in the latter induces a corresponding change in the former have inspired antirealistic views supposedly based on the intrinsic incommensurability of theories. The import of the term "mass", for example, would be so intrinsically relativised to a particular theory that it would be impossible, in consequence, to claim that what special relativity says about the variation of mass with velocity stands in contradiction with Newtonian claims about the invariability of mass. But what is reasonable about the holistic contention, that meaning depends upon the postulates of the theory, does not justify the claim that the import of terms is *uniquely* circumscribed by the formulation of a particular theory as it stands at a given time.

On the other hand, arguments based on general philosophical principles in support of realist views were equally unconvincing. The miracle argument and the scheme of inference to the best explanation on which it seems to rely were unpersuasive. Moreover, the more specific claims about the preservation of reference failed to carry conviction. What Aristotle did and didn't call water can't reasonably be understood as generally coinciding with what we do and don't take to be water, even if he sometimes made claims we would accept. Atomism is an idea that has been around since antiquity, but if it is empirical justification that is to underlie a claim of continuity of reference with the modern use of the term "atom", then those of our predecessors whose use of the term can be considered to have the same reference as ours must date from a considerably more recent period. Arguably, even Dalton's use of the term fails to meet the criterion and it seems we have to jump to the turn of the next century to make the connection. And even if there were things of value that have been saved from caloric and ether theories, that doesn't amount to persisting in referring to ether and caloric.

The rather extreme realist stance that I'm criticising here has been summarised by Stathis Psillos (2018, pp. 24–5) in three theses: first, a non-verificationist view of reference which understands theories literally and acknowledges the world as "populated by unobservable entities". Second, truth is understood as correspondence with reality, in order that it can explain the success of science. Third, the continuity of science is understood as based on the continuity of reference, according to which successive theories are taken to refer to the same entities, even though abandoned theories might have mischaracterised these entities. Such "referential stability in theory-change is indispensable", says Psillos, if conceptions of reference, on which "it becomes inevitable that every time the theory changes, the meanings of *all* terms change, ... and given that reference is supposed to be fixed by descriptions, meaning change is taken to lead [to] reference variance", are to be avoided and continuity saved.

Against this, first, the observation-theory distinction on which the positivists based their instrumentalist thesis has never been satisfactorily established or successfully revamped, and the Duhemian holistic critique undermines this basic component of antirealism. In particular, the important distinction between macroscopic and microscopic realms, mistaken by positivists as a distinction between observable and theoretical and later by certain realists as a distinction between the merely superficial and the deeply theoretical, gives no support to the coherence of the observation-theory distinction. So the realist need only speak of understanding theories literally. Second, the correspondence theory of truth has been roundly criticised as circular (Sect. 3.3 and Asay 2018) and attempts to bolster up the bare claim of truth with talk about corresponding to reality adds nothing of substance. Finally, continuity is not threatened by the scare story about meaning change, as indicated a couple of paragraphs back. In any case, far too much emphasis is placed on reference. Duhem could advocate his continuity thesis without calling upon any such idea. As illustrated by examples of hydrodynamics (Sect. 2.7) and chemical substance (Sect. 7.3.1), progress typically takes the form of the refinement of concepts, as we've seen with pressure and chemical substance, or revising archaic ideas about heat by distinguishing the state of warmth (measured by temperature) and the process of heating, and the introduction of new concepts such as energy and entropy, the development of which can be traced over the years. Trying to construe this as concerned with entities by speaking of the extensions of predicates hasn't worked in the paradigm case of "water", and the kind of continuity of reference that Putnam sought to justify has no more been established than has incommensurability due to holistic meaning change. I suspect that all this talk of entities derives from twentieth-century quests for fundamental particles. Previously, physical ontology was concerned with macroscopic quantities of matter, distinguished as different kinds of matter, to which radiation was subsequently added. Conjectures about a "deeper, underlying" realm of ether, caloric, a spiritual bearer of consciousness and microscopic corpuscles were speculative, failing to gain unanimous backing and lacking any clear experimental support.

Accepting modern, empirically well-justified theories as literally true doesn't seem to call for such extreme realist claims, which can hardly be regarded as necessary components of the basic realist stance. However, a realistic realist stance can't simply amount to accepting the literal truth of theories as currently formulated—not, at any rate, without some qualification and clarification in the face of the real nature of scientific theories. For one thing, theories are typically understood to hold within restricted domains, even if the formalism gives no hint of restrictions. Macroscopic theories, such as irreversible thermodynamics, are understood by chemists and physicists to concern macroscopic objects. Nevertheless, the theory is formulated in terms of the mathematics of the differential and integral calculus, expressing how features such as temperature vary smoothly from one point of space at one instant of time to another point, perhaps at another instant, i.e. continuously whilst maintaining differentiability at each point, as modelled by the real number continuum. But this is inconsistent with the understanding of matter at the microscopic level as consisting of discrete molecular species.

Rash claims are sometimes made, perhaps less frequently in recent times, to the effect that such inconsistency is resolved by dispensing with macroscopic theories and reducing their truth content to purely microscopic theories. Philosophers have in the past sought to build on what they took to be the paradigm case of reduction, namely that of the temperature of a macroscopic body to the average kinetic energy of the collection of molecules, in terms of which the macroscopic body is allegedly eliminated. But, continuing in the spirit of Duhem (Sect. 6.6), the reductionist claim reducing the macroscopic property of temperature to properties of microscopic objects together with the corresponding ontological claim of eliminating the existence of the macroscopic body in favour of the existence of the microscopic objects is highly problematic,[24] throwing doubt on the more sweeping reductionist claim in all its generality. A realist view which avoids inconsistency by appealing to a universally valid microscopic theory to which other theories can be reduced doesn't, therefore, seem to be a scientifically feasible alternative. The entire macroscopic theory is not taken to be literally true as it stands, which would be reflected in an accurate regimentation representing the ontological and theoretical claims that can be sustained.

This is not to denigrate the goal of striving towards a consistent, unified theory, but to recognise that no such theory is now available. A complete theory, it should be said, is one which explicitly deals with everything, including for example geological and biological phenomena. We should be suspicious of any claims to merely lay down guidelines for determining such things "in principle". It must provide the chemist with the detailed description of the properties and inter-reactions of substances, the biologist with the details of plant and animal behaviour and the geologist with details of rock formation. Otherwise, by what criterion would the putative reduction be assessed as adequate? Structural realism, which is motivated by the abstract mathematical structures identified in mathematical physics but entirely absent in other sciences, must rely on the reductionist thesis if it is to have any claim to be generally applicable to science as a whole.

The realist attitude is evident in scientists' specific theoretical concerns. Models are criticised, despite their empirical success or pedagogical convenience, from critical perspectives in the quest for correct explanations, as illustrated at the end of Sect. 7.2. In this spirit, theories are taken as they are understood, not necessarily in all the detail of their standard formulation, with restrictions to particular domains of applicability and interpreted literally only with respect to their considered domain of applicability. Why would scientists distinguish between the theory of

---

[24]There is no general microscopic circumstance underlying or realising all cases in which temperature is displayed (a gas at equilibrium, radiation, a solid undergoing a phase change to liquid, the nuclear spin system of lithium nuclei in a lithium chloride crystal subject to a magnetic field). Further, construing the relation between ideal gas temperature and average kinetic energy of the molecules as reductive is circular (Needham 2009b). Another view of the relation between micro- and the macroscopic theory, echoing Duhem, is that they work in tandem. This is how the relation between gas temperature and the average kinetic energy of the gas molecules should be viewed.

irreversible thermodynamics as formulated and as understood if they didn't take the macroscopic theory as they understand it to be true as opposed to taking the entire formal theory merely as an instrument for making predictions? It is a very common feature of scientific practice to see such a distinction being made between what is held true and what is merely formal machinery to facilitate the working or application, not only in the case of whole theories but more especially in the modelling of specific phenomena. The so-called water drop model of the nucleus, for example, likens the bombardment of a nucleus with high-energy particles to the way surface tension governs the behaviour of drops of water upon impact. Clearly, only certain specific features of the model are taken to be applicable to, i.e. provide true descriptions of, the nature of heavy nuclei (what Levy (2015) calls a partially true description). There is no claim that a nucleus can be literally described by the macroscopic notion of a liquid in all its aspects. Accordingly, the false points of the analogy are not grist to the pessimistic inductivist's mill, and are not reasonably added to the falsity content of a model or theory in the verisimilitude theorist's calculations.

So a moderate realist stance which is feasible in the present state of science steers a course between the extremes of antirealism and realism that have been nurtured by general arguments like the pessimistic induction and the miracle argument, and upholds a patchwork ontology based on the entities in the domains to which empirically well-founded theories are understood to apply. It is guided by specific considerations from the various sciences rather than general philosophical principles applying across the sciences. Chemists believe in the divisions of matter into substances and phases at the macroscopic level and molecular species such as full-blown molecules, ions, and transient entities like free radicles and reaction intermediates at the microscopic level. This has much in common with what biologists accept as the biochemical underpinnings of life, together with organisms, their organs and the populations they constitute. There are planets, stars, galaxies and other denizens of the universe at large. Many things are uncertain—the dark matter postulated to explain gravitational effects observed in the universe, for example, about which we have only a tenuous grasp. But there is no ether and no caloric. Modern physics presents problems of conceptualisation that have motivated antirealist views in the course of the twentieth century. The same sources have inspired structural realism, which pulls in the opposite direction, and has been motivated in more recent times in connection with difficulties of formulating a general notion of a material object applicable to the mysterious nature of the quantum domain and issues arising in uniting this with cosmology. There may well be problems of envisaging the character of molecules in isolation and understanding how the mysterious nature of indistinguishable particles of the quantum domain can constitute distinguishable bodies at some level of organisation. But without the imperative of reduction, these challenges needn't detract from the firm belief in the ontology and ideology underlying what modern science has achieved.

References

Akeroyd, M. (2003). The Lavoisier-Kirwan debate and approaches to the evaluation of theories. *Annals of the New York Academy of Sciences, 988*, 293–301.

Ariew, R. (1984). The Duhem thesis. *British Journal for the Philosophy of Science, 35*, 313–325.

Aristotle (1984). *The complete works of Aristotle* (Vol. 1). Barnes, J. (Ed.). Princeton: Princeton University Press.

Arp, H. C., Furböridge, G., Hoyle, F., Narlikar, J. V., & Wickramasinghe, N. C. (1990). The extragalactic universe: An alternative view. *Nature, 346*, 807–812.

Arrhenius, S. (1885). van t'Hoff, J. H., *Études de dynamique chimique*. Amsterdam: Frederik Muller & C:o, 1884. *Nordisk Revy*, årgång II (1884–1885), nr. 28 (31 mars), 364–365.

Asay, J. (2018). Realism and theories of truth. In J. Saatsi (Ed.), *The Routledge handbook of scientific realism* (pp. 383–393). London: Routledge.

Ayer, A. J. (1946). *Language, truth and logic* (1st ed. 1936; 2nd ed.). London: Victor Gollancz.

Babbage, C. (1830). *Reflections on the decline of science in England*. London: B. Fellows; Reprinted Gregg International, Farnborough, 1969.

Benveniste, J. (1988). Dr. Jacques Benveniste replies. *Nature, 334*, 291.

Bird, A. (2007). What is scientific progress? *Noûs, 41*, 64–89.

Bjereld, U. L. F., Demker, M., & Hinnfors, J. (2002). *Varför vetenskap?* Lund: Studentlitteratur.

Bohm, D. (1996). *The special theory of relativity*. London: Routledge.

Bohr, N. (1965). The structure of the atom. In *Nobel lectures, physics 1922–1941* (pp. 7–43). Amsterdam: Elsevier Publishing Company.

Bogen, J., & Woodward, J. (1988). Saving the phenomena. *Philosophical Review, 97*, 303–352.

Boring, E. G. (1957). *A history of experimental psychology* (2nd ed.). New York: Appleton-Century Crofts.

Braun, S., Ronzheimer, J. P., Schreiber, M., Hodgman, S. S., Rom, T., Bloch, I., & Schneider, U. (2013). Negative absolute temperature for motional degrees of freedom. *Science, 339*, 52–55.

Brink, C. (1989). Verisimilitude: Views and reviews. *History and Philosophy of Logic, 10*, 181–201.

Brush, S. G. (1989). Prediction and theory evaluation: The case of light bending. *Science, 246*, 1124–1129.

Brouzeng, P. (1987). *Duhem: Science et providence*. Paris: Belin.

Bukulich, A. (2008). *Reexamining the quantum-classical relation: Beyond reductionism and pluralism*. Cambridge: Cambridge University Press.

Callen, H. B. (1985). *Thermodynamics and an introduction to thermostatistics*. New York: John Wiley.

© Springer Nature Switzerland AG 2020
P. Needham, *Getting to Know the World Scientifically*, Synthese Library 423,
https://doi.org/10.1007/978-3-030-40216-7

Chakravartty, A. (2008). What you don't know can't hurt you: Realism and the unconceived. *Philosophical Studies, 137*, 149–158.

Chalmers, A. (2017). *One hundred years of pressure: Hydrostatics from Stevin to Newton.* Dordrecht: Springer.

Chang, H. (2003). Preservative realism and its discontents: Revisiting caloric. *Philosophy of Science, 70*, 902–912.

Chang, H. (2012a). Acidity: The persistence of the everyday in the scientific. *Philosophy of Science, 79*, 690–700.

Chang, H. (2012b). Joseph Priestley (1722–1804). In R. F. Hendry, P. Needham, & A. J. Woody (Eds.), *Handbook of the philosophy of science* (Vol. 6, Philosophy of chemistry, pp. 55–62). Amsterdam: Elsevier.

Chomsky, N. (1959), Review of B. F. Skinner. *Verbal Behavior, Language, 35*, 26–58.

Collins, H., & Pinch, T. (1998). *The Golem: What you should know about science* (2nd ed.). Cambridge: Cambridge University Press.

Coulson, C. A. (1955). The contributions of wave mechanics to chemistry. *Journal of the Chemical Society*, 2069–2084.

Cranor, C. F. (1993). *Regulating toxic substances: A philosophy of science and the law.* New York: Oxford University Press.

Dancoff, S. (1952). Does the neutrino really exist? *Bulletin of the Atomic Scientists, 8*, 139–141.

Daub, E. E. (1976, December). Gibbs' phase rule: A centenary retrospect. *Journal of Chemical Education, 53*, 747–751.

Davenas, E., et al. (1988). Human basophil degranulation triggered by very dilute antiserum against IgE. *Nature, 333*, 816–818. See also editorial comment, p. 787 and subsequent discussion, 334, 285–291.

Davy, H. (1840). *Works of Sir Humphrey Davy* (Vol. VII). London: Smith, Elder and Co.

Dellsén, F. (2016). Scientific progress: Knowledge versus understanding. *Studies in History and Philosophy of Science, 56*, 72–83.

Dellsén, F. (2018). Scientific progress, understanding, and knowledge: Reply to Park. *Journal for General Philosophy of Science, 49*(3), 451–459.

Descartes, R. (1637). Discourse on the method of rightly conducting one's reason and seeking the truth in the sciences. In *The philosophical writings of Descartes* (Vol. I, trans: J. Cottingham, R. Stoothoff, & D. Murdoch). Cambridge: Cambridge University Press, 1985.

Descartes, R. (1991). *The philosophical writings of Descartes* (The Correspondence, Vol. III, trans: Cottingham, J., Stoothoff, R., Murdoch, D., & Kenny, A.). Cambridge: Cambridge University Press.

Dorling, J. (1973). Demonstrative induction: Its significant role in the history of physics. *Philosophy of Science, 40*, 360–372.

Douglas, H. (2000). Inductive risk and values in science. *Philosophy of Science, 67*, 559–579.

Douglas, H. (2004). Prediction, explanation, and dioxin biochemistry: Science in public policy. *Foundations of Chemistry, 6*, 49–63.

Drake, S. (1973). Galileo's experimental confirmation of horizontal inertia: Unpublished manuscripts. *Isis, 64*(1973), 291–305.

Drake, S. (1978). *Galileo at work: His scientific biography.* Chicago: University of Chicago Press.

Drake, S., & Kowal, C. T. (1980). Galileo's sighting of Neptune. *Scientific American, 243*, 52–59.

Duhem, H. P. (1936). *Un savant Français: Pierre Duhem.* Paris: Pluon.

Duhem, P. (1886). *Le potentiel thermodynamique et ses applications à la mécanique chimique et à l'étude des phénomènes électriques.* Paris: A. Hermann.

Duhem, P. (1887). Étude sur les travaux thermodynamiqes de M. J. Willard Gibbs. *Bulletin des Sciences Mathmatiques, 11*, 122–148 and 159–176. Translated in Duhem (2011).

Duhem, P. (1888). *Théorie nouvelle de l'aimantation par influence fondée sur la thermodynamique.* Paris: Gauthier-Villars.

Duhem, P. (1891). *Hydrodynamique, elasticité, acoustique* (2 Vols.). Paris: Hermann.

Duhem, P. (1893). Une Nouvelle Théorie du Monde Inorganique. *Revue des questions scientifiques, 33*, 99–133.

Duhem, P. (1894). Commentaire aux principes de la Thermodynamique. Troisième Partie: Les équations générales de la Thermodynamique. *Journal de Mathématiques Pure et Appliquées, 10*, 207–285. Translated in Duhem (2011).

Duhem, P. (1895). Les Théories de la Chaleur. *Revue des Deux Mondes, 129*, 869–901; 130, 380–415, 851–68. Translated as "Theories of Heat" in Duhem (2002).

Duhem, P. (1897). Thermochimie, à propos d'un livre récent de M. Marcelin Berthelot. *Revue des questions scientifiques, 42*, 361–92. Translated in Duhem (2002).

Duhem, P. (1898). La loi des phases, à propos d'un livre récent de M. Wilder D. Bancroft. *Revue des questions scientifiques, 44*, 54–82. Translated in Duhem (2002).

Duhem, P. (1902). *Le mixte et la combinaison chimique: Essai sur l'évolution d'une idée.* Paris: C. Naud. Translated in Duhem (2002).

Duhem, P. (1903a). *L'Evolution de la Mécanique.* Paris: A. Joanin.

Duhem, P. (1903b). *Recherches sur l'Hydrodynamique.* Paris: Gauthier-Villars.

Duhem, P. (1903–1913). *Études sur Léonard de Vinci* (3 Vols.): Première série, ceux qu'il a lus, ceux qui l'ont lu (1906); Deuxième série, ceux qu'il a lus, ceux qui l'ont lu, (1909); Troisième série, Les précurseurs parisiens de Galilée (1913). Paris: A. Hermann.

Duhem, P. (1905–1906). *Les origines de la statique* (2 Vols.). Paris: Hermann.

Duhem, P. (1906a). *Recherches sur l'Élasticité.* Paris: Gauthier-Villars.

Duhem, P. (1906b). *La théorie physique: Son objet – sa structure.* Paris: Chevalier et Rivière; 2nd ed. (text unchanged, two appendices added) 1914; Reprinted Vrin, Paris, 1981. Translated in Duhem (1954).

Duhem, P. (1954). *The aim and structure of physical theories* (trans. of 2nd. ed. of Duhem (1906b) by Philip Wiener). New Jersey: Princeton University Press.

Duhem, P. (2002). *Mixture and chemical combination, and related essays* (trans. and ed.: Needham, P.). Dordrecht: Kluwer.

Duhem, P. (2011). *Commentary on the principles of thermodynamics* (trans. and ed.: Needham, P.). Dordrecht: Springer.

Earley, J. E. (2005). Why there is no salt in the sea. *Foundations of Chemistry, 7*, 85–102.

Earman, J., & Glymour, C. (1980). Relativity and eclipses: The British eclipse expeditions of 1919 and their predecesssors. *Historical Studies in the Physical Sciences, 11*, 49–85.

Einstein, A. (1905 [1956]). Über die von der molekularkinetischen Theorie der Wärme geforderte Bewegung von in ruhenden Flüssigkeiten suspendierten Teilchen. *Annalen der Physik, 17*(8), 549–560; translated in *Investigations on the theory of the Brownian movement.* New York: Dover.

Fernández Moreno, L. (2001). Tarskian truth and the correspondence theory. *Synthese, 126*(1–2), 123–147.

Feyerabend, P. K. (1962). Explanation, reduction, and empiricism. In H. Feigl & G. Maxwell (Eds.), *Minnesota studies in the philosophy of science* (Vol. 3, pp. 28–97). Minneapolis: University of Minnesota Press.

Feynman, R. P. (1939). Forces in molecules. *Physical Review, 56*, 340–343.

Frank, S. T. (1973). Aural sign of coronary-artery disease. *New England Journal of Medicine, 289*, 327–328.

Franklin. A. (1986). *The neglect of experiment.* Cambridge: Cambridge University Press.

Frege, G. (1918 [1977]). Thoughts (trans: Geach, P. T. of "Der Gedanke"). In *Logical investigations.* Oxford: Blackwell.

Frické, M. (1976). The rejection of Avogadro's hypotheses. In C. Howson (Ed.), *Method and appraisal in the physical sciences: The critical background to modern science 1800–1905* (pp. 277–307). Cambridge: Cambridge University Press.

Galileo, G. (1613 [1957]). *Letters on sunspots*, trans.: Drake, S., and C. D. O'Malley, Ed., *The Controversy on the Comets of 1618.* Philadelphia: University of Pennsylvania Press, 1960. Extracts in *Discoveries and Opinions of Galileo*, Ed. and trans. S. Drake. New York: Doubleday Anchor Books.

Galileo, G. (1632 [1967]). *Dialogues concerning the two chief world systems* (trans: Drake, S.). Berkeley and Los Angeles: University of California Press.

Galileo, G. (1638 [1954]). *Dialogue concerning two new sciences* (trans: Crew, H., de Savio, A.). New York: Dover.

Giedymin, J. (1982). *Science and convention.* Oxford: Pergamon Press.

Gettier, E. (1964). Is justified true belief knowledge? *Analysis, 23,* 121–123.

Griffiths, D. (2004). *Introduction to elementary particles.* New York: Wiley.

Harman, G. (1965). Inference to the best explanation. *Philosophical Review, 74,* 88–95.

Heitler, W., & London, F. (1927). Wechselwirkung neutraler Atome und homöopolare Bindung nach der Quantenmechanik. *Zeitschrift für Physik, 44,* 455–472. Trans. in Hettema, H. (2000). *Quantum chemistry: Classic scientific papers* (pp. 140–55). Singapore: World Scientific.

Hempel, C. G. (1954 [1965]). A logical appraisal of operationalism. *Scientific Monthly, 79,* 215–220. Reprinted in *Aspects of scientific explanation.* Toronto: Free Press, Collier-Macmillan.

Hempel, C. G. (1950). Problems and changes in the empiricist criterion of meaning. *Revue Internationale de Philosophie, 4*(11), 41–63; amplified in Hempel (1965).

Hempel, C. G. (1965). Science and human values. In *Aspects of scientific explanation* (pp. 81–96). Toronto: Free Press, Collier-Macmillan.

Hilbert, D. (1899 [1971]). *Grundlagen der Geometrie.* Leipzig: Teubner. Translated from the 10th ed. by Leo Unger as *Foundations of geometry.* La Salle: Open Court, 1971.

Hoefer, C., & Rosenberg, A. (1994). Empirical equivalence, underdetermination, and systems of the world. *Philosophy of Science, 61,* 592–607.

Hume, D. (1739 [1967]). *A treatise of human nature* L. A. Selby-Bigge (Ed.). Oxford: Clarendon Press.

Hughes, I., & Hase, T. (2010). *Measurements and their uncertainties: A practical guide to modern error analysis.* Oxford: Oxford University Press.

Hunt, B. J. (1991). *The Maxwellians.* New York: Cornell University Press.

Iliopoulos, J. (2017). *The origin of mass: Elementary particles and fundamental symmetries.* Oxford: Oxford University Press.

Jaki, S. L. (1984). *Uneasy genius: The life and work of Pierre Duhem.* The Hague: Martinus Nijhoff.

Kant, I. (1783 [1966]). *Prolegomena.* Riga: Hartknoch; Eng. trans. by Peter G. Lucas, Manchester University Press.

Kim, J. (2005). *Physicalism, or something near enough.* Princeton: Princeton University Presss.

Kirvan, R. (1789). *An essay on phlogiston and the constitution of acids: A new edition to which are added, notes, exhibiting and defending the antiphlogistic theory; and annexed to the French edition of this work; by Messrs de Morveau, Lavoisier, de la Place, Monge, Berthollet, and de Fourcroy,* translated into English, with additional remarks and replies by the author, J. Johnson, London (Facsimile edition published in 1968 by F. Cass, London.).

Kitcher, P. (2001). *Science, truth and democracy.* New York: Oxford University Press.

Kjeldstadli, K. (1998). *Det förflutna är inte vad det en gång var.* Lund: Studentlitteratur.

Klein, U. (1994). Origin of the concept of chemical compound. *Science in Context, 7,* 163–204.

Klotz, I. M. (1980). The N-ray affair. *Scientific American, 242,* 122–131.

Knight, D. (1978). *The transcendental part of chemistry.* Folkestone: Dawson.

Kociba, R., Keyes, D. G., Beyer, J. E., Carreon, R. M., Wade, C. E., Dittenber, D. A., Kalnins, R. P., Frauson, L. E., Park, C. N., Barnard, S. D., Hummel, R. A., & Humiston, C. G. (1978). Results of a two-year chronic toxicity and oncogenicity study of 2,3,7,8-tetracholorodibenzo-p-dioxin in rats. *Toxicology and Applied Pharmacology, 46,* 279–303.

Kuhn, T. S. (1957). *The Copernican revolution.* Cambridge: Harvard University Press.

Kuhn, T. S. (1962 [1970]). *The structure of scientific revolutions* (2nd ed.). Chicago: University of Chicago Press.

Kuhn, T. S. (1978). *Black-body theory and the quantum discontinuity 1894–1912.* New York: Oxford University Press.

Kutzelnigg, W. (1996). Friedrich Hund and chemistry. *Angewandte Chemie, International Edition in English, 35,* 573–586.

Ladyman, J. (2000). What's really wrong with constructive empiricism? van Fraassen and the Metaphysics of Modality. *British Journal for the Philosophy of Science, 51,* 837–856.

Ladyman, J. (2004). Constructive empiricism and modal metaphysics: A reply to Monton and van Fraassen. *British Journal for the Philosophy of Science, 55*, 755–765.

Ladyman, J. (2009). Structural realism. In E. N. Zalta (Ed.), *Stanford encyclopedia of philosophy* (Summer 2009 Edition). http://plato.stanford.edu/entries/structural-realism/

Lakatos, I. (1970). Falsification and the methodology of scientific research programmes. In I. Lakatos & A. Musgrave (Eds.), *Criticism and the growth of knowledge*. Cambridge: Cambridge University Press.

Lakatos, I. (1978). *The methodology of scientific research programmes* (Philosophical papers, Vol. 1). Cambridge: Cambridge University Press.

Lang, S. (1998). *Challenges*. New York: Springer.

Laudan, L. (1981 [1984]). A confutation of convergent realism. *Philosophy of Science, 48*, 19–49. Reprinted in J. Leplin (Ed.), *Scientific realism* (pp. 218–249). Berkeley and Los Angeles: University of California Press.

Lavoisier, A. (1789 [1965]). *Traité Élémentaire de Chimie*. Paris. Trans. by Robert Kerr (1790) as *Elements of chemistry*. New York: Dover. Reprint.

Levy, A. (2015). Modeling without models. *Philosophical Studies, 172*(3), 781–798.

Lewis, G. N. (1916). The atom and the molecule. *Journal of the American Chemical Society, 38*, 762–785.

Lewis, G. N. (1917). The static atom. *Science, 46*, 297–302.

Lewis, G. N. (1923). *Valence and the structure of atoms and molecules*. New York: Chemical Catalog Co.; Reprinted in 1966 by Dover, New York.

Lubbe, S. C. C. van der, & Fonseca Guerra, C. (2019). The nature of hydrogen bonds: A delineation of the role of different energy components on hydrogen bond strengths and lengths. *Chemistry—An Asian Journal, 14*(16), 2760–2769.

Lubbe, S. C. C. van der, Zaccaria, F., Sun, X., & Fonseca Guerra, C. (2019). Secondary electrostatic interaction model revised: Prediction comes mainly from measuring charge accumulation in hydrogen-bonded monomers. *Journal of the American Chemical Society, 141*, 4878–4885.

Maddox, J., Randi, J., & Stewart, W. W. (1988). 'High-dilution' experiments a delusion. *Nature, 334*, 287–290.

Maier, A. (1982). *On the threshold of exact science: Selected writings of Anneliese Maier on late medieval natural philosophy*. Philadelphia: University of Pennsylvania Press.

Masterman, M. (1970). The nature of a paradigm. In I. Lakatos & A. Musgrave (Eds.), *Criticism and the growth of knowledge* (pp. 59–89). Cambridge: Cambridge University Press.

Maxwell, J. C. (1875 [1890]). Atom. *Encyclopedia Britannica* (9th ed., Vol. III, pp. 36–49). Reprinted in W. D. Niven (Ed.), *The collected scientific papers of James Clerk Maxwell* (Vol. II, pp. 445–484). Cambridge: Cambridge University Press.

McMullin, E. (1983). Values in science. In P. D. Asquith & T. Nickles (Eds.), *Proceedings of the 1982 biennial meeting of the philosophy of science association* (Vol. 2, pp. 3–28). East Lansing: Philosophy of Science Association.

Metzger, H., & Dreskin, S. (1988). Only the smile is left. *Nature, 334*, 375. See also editorial comment, p. 367.

Miller, D. G. (1966). Ignored intellect: Pierre Duhem. *Physics Today, 19*, 47–53.

Miller, D. G. (1971). Duhem, Pierre-Maurice-Marie. In C. C. Gillispie (Ed.), *Dictionary of scientific biography* (Vol. IV, pp. 225–233). New York: Scribner.

Miller, D. W. (1974). Popper's qualitative theory of verisimilitude. *British Journal for the Philosophy of Science, 25*, 166–177.

Millikan, R. A. (1913). On the elementary electrical charge and the Avogadro constant. *Physical Review, 2*, 109–143.

Mizrahi, M. (2013). The pessimistic induction: A bad argument gone too far. *Synthese, 190*, 3209–3226.

Mizrahi, M. (2018). The 'positive argument' for constructive empiricism and inference to the best explanation. *Journal for General Philosophy of Science, 49*(3), 461–466.

Mulliken, R. S. (1931). Bonding power of electrons and theory of valence. *Chemical Reviews, 9*, 347–388.

Nash, L. K. (1957). The atomic molecular theory. In J. B. Conant & L. K. Nash (Eds.), *Harvard case histories in experimental science* (pp. 217–321). Cambridge: Harvard University Press.

Naylor, R. (1974). Galileo's simple pendulum. *Physis, 16*(1974), 32–46.

Needham, P. (1998). Duhem's physicalism. *Studies in History and Philosophy of Science, 29*, 33–62.

Needham, P. (2000). Duhem and Quine. *Dialectica, 54*, 109–132.

Needham, P. (2004). When did atoms begin to do any explanatory work in chemistry? *International Studies in the Philosophy of Science, 8*, 199–219.

Needham, P. (2008a). Is water a mixture?—Bridging the distinction between physical and chemical properties. *Studies in History and Philosophy of Science, 39*, 66–77.

Needham, P. (2008b). Resisting chemical atomism: Duhem's argument. *Philosophy of Science, 75*, 921–931.

Needham, P. (2009a). An Aristotelian theory of chemical substance. *Logical Analysis and History of Philosophy, 12*, 149–164.

Needham, P. (2009b). Reduction and emergence: A critique of Kim. *Philosophical Studies, 146*, 93–116.

Needham, P. (2010). Substance and time. *British Journal for the Philosophy of Science, 61*(2010), 485–512.

Needham, P. (2012). Compounds and Mixtures. In R. F. Hendry, P. Needham, & A. J. Woody (Eds.), *Handbook of the philosophy of science* (Vol. 6, Philosophy of chemistry, pp. 271–290). Amsterdam: Elsevier.

Needham, P. (2013). Hydrogen bonding: Homing in on a tricky chemical concept. *Studies in History and Philosophy of Science, 44*, 51–66.

Needham, P. (2014). The source of chemical bonding. *Studies in History and Philosophy of Science, 45C*, 1–13.

Needham, P. (2018). Scientific realism and chemistry. In J. Saatsi (Ed.), *The Routledge handbook of scientific realism* (pp. 345–356). London: Routledge.

Newman, W. R. (1996). The alchemical sources of Robert Boyle's corpuscular philosophy. *Annals of Science, 53*, 567–585.

Newton, I. (1687 [1999]). *The principia: Mathematical principles of natural philosophy* (trans: Bernard Cohen, I., Whitman, A.). Berkely: University of California Press.

Norton, J. (1993). The determination of theory by evidence: The case for quantum discontinuity, 1900–1915. *Synthese, 97*, 1–31.

Norton, J. (1994). Science and certainty. *Synthese, 99*, 3–22.

Norton, J. (2003). A material theory of induction. *Philosophy of Science, 70*, 647–670.

Norton, J. (2006). The formal equivalence of grue and green and how it undoes the new riddle of induction. *Synthese, 150*, 185–207.

Norton, J. (2010). There are no universal rules for induction. *Philosophy of Science, 77*, 765–777.

Norton, J. (2011). History of science and the material theory of induction: Einstein's quanta, mercury's perihelion. *European Journal for Philosophy of Science, 1*, 3–27.

Norton, J. (2014). A material dissolution of the problem of induction. *Synthese, 191*, 671–690.

Norton, J. (2015). Replicability of experiment. *Theoria: An International Journal for Theory, History and Foundations of Science, 30*(2), 229–248.

Nye, M. J. (1986). *Science in the provinces: Scientific communities and provincial leadership in France, 1860–1930*. Berkeley: University of California Press.

Nye, M. J. (1993). *From chemical philosophy to theoretical chemistry: Dynamics of matter and dynamics of disciplines 1800–1950*. Berkeley: California University Press.

Ohanian, H. C. (1988). *Classical electrodynamics*. Boston: Allyn and Bacon.

Papineau, D. (1996). Editor's "Introduction". In *The philosophy of science*. Oxford: Oxford University Press.

Park, S. (2017). Does scientific progress consist in increasing knowledge or understanding? *Journal for General Philosophy of Science, 48*(4), 569–579.

Partington, J. R. (1962). *A history of chemistry* (Vol. 3). London: Macmillan.

Pinkava, J., (Ed.). (1987). *The correspondence of the Czech chemist František Wald with W. Ostwald, E. Mach, P. Duhem, J. W. Gibbs and other scientists of that time*. Praha: Academia Nakladatelstvi Ceskoslovenske Akademie Ved.

Poincaré, H. (1902 [1952]). *La Science et l'Hypothèse*. Flammarion, 1902; trans. *Science and hypothesis*. New York: Dover.

Popper, K. R. (1968). *The logic of scientific discovery*, trans. of *Logik der Forschung*, 1934, with additional appendices. London: Hutchinson.

Popper, K. R. (1969). *Conjectures and refutations* (3rd ed.). London: Routledge and Kegan Paul.

Popper, K. R. (1983). *A pocket Popper*, D. Miller (Ed.). Oxford: Fontana.

Price, D. J. De S. (1959). Contra Copernicus: A critical re-estimation of the mathematical planetary theory of Ptolemy, Copernicus and Kepler. In M. Claggett (Ed.), *Critical problems in the history of science*. Wisconsin: University of Wisconsin Press.

Psillos, S. (1999). *Scientific realism: How science tracks truth*. London: Routledge.

Psillos, S. (2018). The realist turn in the philosophy of science. In J. Saatsi (Ed.), *The Routledge handbook of scientific realism* (pp. 20–34). London: Routledge.

Putnam, H. (1962). The analytic and the synthetic. In H. Feigl & G. Maxwell (Eds.), *Minnesota studies in the philosophy of science* (Vol. III). Minnesota: University of Minnesota Press; Reprinted in Putnam (1975b).

Putnam, H. (1975a). *Mathematics, matter and method* (Philosophical papers, Vol. 1). Cambridge: Cambridge University Press.

Putnam, H. (1975b). *Mind, language and reality* (Philosophical papers, Vol. 2). Cambridge: Cambridge University Press.

Quine, W. V. (1936). Truth by convention. In O. H. Lee (Ed.), *Philosophical essays for A. N. Whitehead*. London: Longmans; Reprinted in *The ways of paradox* (2nd ed.). Cambridge: Harvard University Press, 1976.

Quine, W. V. (1951). Two dogmas of empiricism. *Philosophical Review, 60*; Reprinted with changes in *From a logical point of view* (rev ed.). Cambridge: Harvard University Press, 1961.

Quine, W. V. (1960). *Word and object*. Cambridge: M.I.T. Press.

Quine, W. V. (1974). On Popper's negative methodology. In P. A. Schilpp (Ed.), *The philosophy of Karl Popper*. Illinois: Open Court.

Quine, W. V. (1975). On empirically equivalent systems of the world. *Erkenntnis, 9*, 313–328.

Quine, W. V. (1976). A comment on Grünbaum's claim. In S. Harding (Ed.), *Can theories be refuted?*. Dordrecht: Reidel.

Quine, W. V. (1996). Progress on two fronts. *Journal of Philosophy, 93*, 159–163.

Quine, W. V., & Ullian, J. S. (1978). *The web of belief* (2nd ed.). New York: Random House.

Raftopoulos, A. (1999). Newton's experimental proofs as eliminative reasoning. *Erkenntnis, 50*(1), 95–125.

Ramsey, N. F. (1956). Thermodynamics and statistical mechanics at negative absolute temperatures. *Physical Review, 103*, 20–28.

Rudner, R. (1953). The scientist qua scientist makes value judgments. *Philosophy of Science, 20*, 1–6.

Russell, B. (1912). *The problems of philosophy*. Home University Library, since reprinted many times by Oxford University Press.

Schupbach, J. N. (2017). Inference to the best explanation, cleaned up and made respectable. In K. McCain & T. Poston (Eds.), *Best explanations* (pp. 39–61). Oxford: Oxford University Press.

Schurz, G., & Weingartner, P. (1987). Verisimilitude defined by relevant consequence-elements. In T. A. F. Kuipers (Ed.), *What is closer-to-the-truth?*. Amsterdam: Rodopi.

Sengers, J. L. (2002). *How fluids unmix: Discoveries by the school of Van der Waals and Kamerlingh Onnes*. Amsterdam: Koninklijke Nederlandse Adakamie van Wetenschappen.

Settle, T. B. (1961). An experiment in the history of science. *Science, 133*, 19–23.

Sklar, L. (1993). *Physics and chance: Philosophical issues in the foundations of statistical mechanics*. Cambridge: Cambridge University Press.

Sobel, D. (1995). *Longitude: The true story of a lone genius who solved the greatest scientific problem of his time*. New York: Walker & Company.

Stanford, P. K. (2018). Unconceived alternatives and the strategy of historical ostension. In J. Saatsi (Ed.), *The Routledge handbook of scientific realism* (pp. 212–224). London: Routledge.

Tichy, P. (1974). On Popper's definition of verisimilitude. *British Journal for the Philosophy of Science, 25*, 155–160.

van Fraassen, B. (1980). *The image of science*. Oxford: Clarendon Press.

van Fraassen, B. (1989). *Laws and symmetry*. Oxford: Clarendon Press.

Wachbroit, R. (1986). Progress: Metaphysical and otherwise. *Philosophy of Science, 53*, 354–371.

Weintraub, R. (2017). Scepticism about inference to the best explanation. In K. McCain & T. Poston (Eds.), *Best explanations* (pp. 170–202). Oxford: Oxford University Press.

Weisberg, M. (2007). Three kinds of idealization. *Journal of Philosophy, 104*, 639–659.

Westfall, R. S. (1980). *Never at rest: A biography of Isaac Newton*. Cambridge: Cambridge University Press.

White, H. O. (1938). *Plagiarism and imitation during the English renaissance*. Cambridge: Harvard University Press.

Whitehead, A. N., & Russell, B. (1910–1912). *Principia mathematica* (3 Vols.); abridged version of 2nd edition, *Principia mathematica to *56* (1967). Cambridge: Cambridge University Press.

Wikforss, Å. (2017). *Alternativa fakta: Om kunskapen och dess fiender*. Stockholm: Fri Tanke förlag.

Williamson, T. (2000). *Knowledge and its limits*. Oxford: Oxford University Press.

Wilson, C. A. (1972). How did Kepler discover his first two laws? *Scientific American, 226*, 92–106.

Wood, R. W. (1904). The n-rays. *Nature, 70*, 530–531.

Worrall, J. (1982). Scientific realism and scientific change. *Philosophical Quarterly, 32*(1982), 201–231.

Worrall, J. (1989). Structural realism: The best of both worlds? *Dialectica, 43*, 99–124.

Wray, K. B. (2018). The atomic number revolution in chemistry: A Kuhnian analysis. *Foundations of Chemistry, 20*, 209–217.

Zahar, E. G. (1983). The Popper-Lakatos controversy in the light of 'Die Beiden Grundprobleme der Erkenntnistheorie'. *British Journal for the Philosophy of Science, 34*, 149–171.

Zupko, J. (2008). John Buridan. *The Stanford Encyclopedia of Philosophy* (Fall 2008 ed.). E. N. Zalta (Ed.), http://plato.stanford.edu/archives/fall2008/entries/buridan/

Index

© Springer Nature Switzerland AG 2020
P. Needham, *Getting to Know the World Scientifically*, Synthese Library 423,
https://doi.org/10.1007/978-3-030-40216-7

Printed in the United States
by Baker & Taylor Publisher Services